SCIENCE FOR LIFE

A MANUAL FOR BETTER LIVING

BRIAN CLEGG

ICON

This edition published in the UK in 2016 by
Icon Books Ltd, Omnibus Business Centre,
39–41 North Road, London N7 9DP
email: info@iconbooks.com
www.iconbooks.com

First edition published in hardback in the UK in 2015 by Icon Books Ltd

Sold in the UK, Europe and Asia
by Faber & Faber Ltd, Bloomsbury House,
74–77 Great Russell Street,
London WC1B 3DA or their agents

Distributed in the UK, Europe and Asia
by TBS Ltd, TBS Distribution Centre, Colchester Road,
Frating Green, Colchester CO7 7DW

Distributed in the USA by
Publishers Group West
1700 4th St.
Berkeley, CA, 94710

Distributed in Australia and New Zealand
by Allen & Unwin Pty Ltd,
PO Box 8500, 83 Alexander Street,
Crows Nest, NSW 2065

Distributed in South Africa by
Jonathan Ball, Office B4, The District,
41 Sir Lowry Road, Woodstock 7925

Distributed in Canada by Publishers Group Canada,
76 Stafford Street, Unit 300
Toronto, Ontario M6J 2S1

ISBN: 978-178578-025-7

Typeset in Minion by Marie Doherty

Printed and bound in the UK by Clays Ltd, St Ives plc

ABOUT THE AUTHOR

Science writer Brian Clegg studied physics at Cambridge University and specialises in making the strangest aspects of the universe – from infinity to time travel and quantum theory – accessible to the general reader. He is editor of www.popularscience.co.uk and a Fellow of the Royal Society of Arts. His previous books include *Inflight Science, Build Your Own Time Machine, The Universe Inside You, Dice World, The Quantum Age* and *Introducing Infinity: A Graphic Guide.*

www.brianclegg.net

CONTENTS

• • •

DIET   1

• • •

EXERCISE   119

• • •

BRAIN   149

PSYCHOLOGY 211

• • •

HEALTH 251

• • •

ENVIRONMENT   329

• • •

FUN   379

INTRODUCTION

Science plays a fundamental role in everyday lives – improving health, increasing life expectancy, enhancing life experience. Yet it can be difficult to get a practical picture of what is really best for us – what's needed is a science 'recipe book' for you and your family. *Science for Life* is exactly that – it presents the best of current scientific advice, cutting through the vested interests and confusing, contradictory statements to give a clear picture of what science is telling us right now about changing our lives for the better.

Part of the problem we face is that we are bombarded in the press by claims that a new substance helps us lose weight or reduces the risk of cancer ... only to find a few weeks later that the same newspaper or magazine says that the same substance is bad for us. This isn't scientists being confused, but rather the media misusing information.

The news media are desperate to grab our attention. So even when they are entirely aware that a claim for, say, a new product is worthless, they will typically plaster the headlines with the claim as if it were true and only later reveal that it has no merit. The trouble is that, by then, many of us will have given up reading.

A great example emerged while this book was being written – the 'drinkable sunscreen' story (see **Drinkable sunscreen** in the **Health** section for more details). Several newspapers in the UK splashed this story. Even the usually responsible *Daily Telegraph* had an opening paragraph reading: 'The days of asking a friend to rub suncream on your back or waiting for your lotion to "sink in" to avoid a sandy situation could be numbered, as a US skincare company claim to have created a drinkable suncream.'

After giving us details of the product and the company making it, the article does bring in experts to say that the product doesn't make sense – but by then it is too late. Admittedly the *Telegraph*'s headline warns that 'experts say it is a gimmick', but we get no such suggestion from the *Daily Mail*, where a travel reporter (not a science reporter) gives us the headline: 'World's first drinkable sun cream goes on sale – and just a teaspoon will offer three hours' protection.' Well, no, it won't – but *Mail* readers may well be convinced it will.

Beyond this kind of marketing 'news', when the results of a scientific trial are published, the newspapers are even more likely to blare out the trial's findings as fact. But there are two problems with this. One is that not all trials are equal. To be useful, a trial has to be properly managed with a large enough number of participants to iron out any statistical oddities, using sensible techniques and ensuring that neither the participants in the trial nor the testers know who is getting the substance being tried out, and who has a harmless substance with no effect (a placebo) to make a comparison. Such trials are called 'double blind' and unless this is done, it has been shown time and again that the expectations of both participants and testers will influence the results.

By comparison, a lot of the 'trials' and 'studies' reported in the press have very small numbers of participants (say ten to twenty), are often based on participants' descriptions of how they feel, and are not properly controlled for error. They may also be run by individuals with a vested interest in the outcome.

Another problem is that one trial is hardly ever enough. Even the best of scientists can make errors, and one of the checks and balances of science is that before a theory or treatment is considered worthwhile, the results have to be duplicated by other scientists and laboratories. All the best results are supported by a number of trials, and the best medical and dietary results typically come from a study that pulls together the results of many trials and combines them, giving more weighting to the best studies. Such 'meta analyses', like those provided by the Cochrane Collaboration, an organisation that specialises in collecting the best evidence on medical issues, are the gold standards in medicine.

Partly as a result of taking time to undertake a wide range of trials, it is also the case that scientists do sometimes change their minds, as they have, for instance, about the health risks from cholesterol in eggs or the chances of saturated fat in your diet giving you heart disease. This is because real science isn't the same as the cartoon caricature version most of us have in our minds.

We tend to think of science as a search for the truth that will result in discovering the fundamental laws of nature and absolute results. In reality it is hardly ever like that. It's just about possible for science to come up with a solid result in a straightforward piece of physics, like Newton's laws of motion – though even those have had to be modified to deal with Einstein's special relativity. But it's quite different when dealing with a complex system like the human body, and how different aspects of life influence it.

One problem is separating out a particular cause. It can be quite difficult, for instance, to say whether people who are obese are more prone to heart attacks because of their diets, because of their lack of exercise or because of the changes to their bodies brought about by obesity.

So what science usually provides us with is the best current agreed position, given the data we have at the moment. That position may, and sometimes will, change when we have better data – as it did over saturated fat. But there is no sensible reason for going with anything other than the best current agreed position until we have further information.

Unfortunately the media often ignore this, which is why, for instance, they made such a huge mistake in publicising Andrew Wakefield's notoriously bad results linking the MMR vaccination to autism in the UK. Wakefield was a single voice, not particularly well-qualified for the task, and based his assertion on a tiny amount of data from a very small, badly selected group of children. Ranked against him was practically every medical expert, with vast numbers of trials showing the opposite conclusion – and yet the media ran with the Wakefield scare story, many parents stopped their children having MMR and the result was large-scale measles outbreaks, causing serious illness and deaths.

Occasionally there will be high-level confrontations, such as the one over statins, the cholesterol-lowering medication, where there is a considerable argument among experts over the exact balance between a definite reduction in risk of heart attack and stroke, and the side effects associated with the drug. But even here, no one with expertise in the field is suggesting we abandon statins – the discussion is over exactly what level of risk makes it worth taking them.

Just to add to the confusion, many of us expect to get information quickly and accurately from the internet – and often the web can be a good source. But there are all too many websites out there that peddle half-truths or downright lies in order to sell a product.

When I was researching one of the topics in this book, **Raspberry ketones** (see page 117), I came across a web page that seemed to give academic credibility to this weight loss remedy, much touted on the internet. The site, which had the look of a glossy women's magazine, and a name that suggested it was run by a newspaper, said: 'We asked the National Centre for Biotechnology at the University of Reading. They confirmed that raspberry ketones fight obesity and increase metabolism.' I thought this was worth checking.

A quick search online suggested there was no 'National Centre for Biotechnology' at the university, but there was a 'National Centre for Biotechnology Education'. Their response was simple: this has nothing to do with us and we will be contacting our legal team. As it happened, in a perhaps unusual burst of honesty, this particular website had the most hilarious disclaimer I have ever seen. At the bottom of the page it said:

> I UNDERSTAND THIS WEBSITE IS ONLY ILLUSTRATIVE OF WHAT MIGHT BE ACHIEVABLE FROM USING THIS/ THESE PRODUCTS, AND THAT THE STORY/COMMENTS DEPICTED ABOVE IS NOT TO BE TAKEN LITERALLY AND SHOULD NOT BE TREATED AS NON-FICTION.

No, your eyes don't deceive you. It was saying that its advice should be considered fictional. But how many of us check at the bottom for such a disclaimer? And it is all too easy for sellers of mumbo jumbo and quack remedies to make fictional references to universities, which we don't have time to follow up. I have spent hours sifting through online claims for health products and all too often they resort to tricks like these.

Other sites (and plenty of celebrity nutritionists' books) have stacks of references, linking their products and ideas to real scientific research. And these look impressive. But what some of them are relying on is that readers are not going to have the time or inclination to check up on these references. When I have done so, many such references are fictional or, where they do refer to a real piece of research, the conclusions of the research are totally different to those claimed in the article. One of the benefits of this book is that I have checked out the truth behind these claims so you don't have to.

Life is too short to always be confused by the latest fads and misdirection from those who are trying to sell you something. The whole point of *Science for Life* is to make the picture clear and simple for you, improving your life choices without compromising the message of the latest scientific research.

AND THERE'S MORE ...

Science for Life isn't complete. There's an important reason for that. Science (and for that matter, life) is ever-changing, and there will always be new and interesting items to add. To make this possible we have set up **www.scienceforlife.info** – this site features both updates for existing articles and new posts on the latest science information that impacts our lives.

There's also the opportunity to drop us an email with new topics that you would like covered. Take a look at **www.scienceforlife.info**.

DIET

• •

Anyone writing a diet book (and that seems to be every celebrity and nutritionist you can name) faces a problem – there is absolutely no rocket science involved. In the case of some diet books, there's no science at all. The embarrassing fact is that all the dietary advice you ever need could be fitted comfortably into a single paragraph. Here we go:

> Don't eat too much – if you are putting on weight, eat less. Eat plenty of fruit and vegetables (and don't make them into smoothies as this ruins the valuable fibre). Make around one third of your diet starchy foods, preferably wholegrain. Don't eat too much processed meat. Drink alcohol moderately, if you must. Avoid sugar and salt as much as possible and don't go overboard on fats, avoiding trans fats entirely. It's not strictly part of a diet, but add 'don't smoke' and 'take sensible exercise' and you've got an instant health plan.

That's it – that's what you pay hard-earned cash to get a diet book for. Every other page in such a book is padding. If you write a diet book you have to find some way to make yours different from the rest. Some do this by straying away from what is most beneficial to include mystical mumbo jumbo. Others find different ways to expand those basics to fill a whole volume by adding lots of rules, or filling it out with healthy recipes (which is fair enough). But that single paragraph is all the diet book you'll ever need.

The problem is that it is very natural to want a quick fix, to hope for a magic wand we can wave to improve our health. But there is very strong evidence that short-term diets do no good whatsoever.

It's far better to make small changes that you carry through from year to year than to go on a crash diet, returning to overdoing it a few weeks later.

In part because of this desire for a quick fix, there are always new miracle foods and dietary wonders that are splashed across the newspapers. The reason this section is a lot more than just that single paragraph is partly to clarify the value of all these different suggested wonder foods and drinks. Some are just nonsense. Others have a grain of truth behind them and are worth considering – but even these aren't magic bullets. Keep coming back to my core paragraph and you can't go too far wrong.

A major problem that we all face in trying to work out what's good for our health is that it is very difficult to spot whether eating a particular thing *is* good or bad for you. This is why we get so many reports in newspapers telling us that something is good or bad for our health. This is also why we were told for so long that saturated fats were worse for us than unsaturated fats, where this no longer seems to be the case.

The problem is that, unlike testing a medication, it is very difficult to do a proper, scientific blind-controlled trial on what we eat. Most dietary studies are observational – they tell us, for instance, that people in the Mediterranean suffer from less heart disease than people in Glasgow. We can also observe that these people have a different kind of diet. But it's hard to say for certain that it is the diet that is giving the benefit – and even harder to identify a particular aspect of the diet, like olive oil or tomatoes – because there are so many other differences between life in the Mediterranean and life in Glasgow, and we don't know what the actual cause is, merely that people in one environment, with their typical lifestyle, are healthier in this aspect than people in the other.

For instance, in 2001 an Australian study was portrayed in the media as showing that people who consume more olive oil get fewer wrinkles. So journalists (often without a science background) got all excited, telling us that consuming olive oil is good for your skin. But the study was not done by taking two similar groups of people and feeding one olive oil while the other received another oil, with neither the people involved nor the scientists knowing which was which, as would be the case in a proper double blind trial. Such a trial on a big enough sample of people over a long enough period would, indeed, show whether eating olive oil helps reduce wrinkles.

Instead, what the trial did was to bring together information on different groups of people from widely varying backgrounds – Australians, Swedes and Greeks, for instance – and find that the level of wrinkling they experienced corresponded reasonably well to the level of olive oil in their diets. But to deduce that the oil reduced wrinkles is to fall for the oldest statistical error in the book – that correlation (where two things vary in a similar fashion) is the same as causation (saying that one causes the other). See the section on **Paracetamol and childhood asthma**, which explores causality and correlation (page 313), for more on this.

In practice, all manner of other differences were likely to be common in the olive oil and wrinkles study. For instance, dietary variations often relate to levels of income, education, living conditions, environment, stress levels, moisturiser use, sleeping patterns and many other things that could have been responsible for the lower levels of wrinkles. To make the assumption that the consumption of olive oil caused the reduction in wrinkles makes no sense. You could almost certainly find some other factor (say reading newspapers) that also varied with the wrinkle levels.

A lot of the media stories seem to be about a particular food or drink (red wine, say, as this is the favourite substance) causing or preventing cancer. It can seem baffling that the same thing potentially has both effects, and it's easy to think that science has got it wrong. But in reality it is the *interpretation* put on the science by journalists and nutritionists (who often don't have proper scientific training) that is at fault. We've already seen one way this can be the case with observational studies. Comparing populations who do and don't drink red wine is fraught with difficulties in determining what causes differences in health. But there's another problem.

When it's claimed that something causes or cures cancer, for instance, what is often the case is that someone has either tested the substance on cells in a laboratory or fed it in large doses to rats and observed the outcome. This can contribute to very valuable research, leading to proper testing of the active chemicals in a way that will see if there are real benefits. But almost all the trials quoted this way *don't* show that the substance being tested, when consumed, will have that effect on cells in the human body. As *Bad Science* author Ben Goldacre points out: 'Fairy Liquid will kill cells in a test tube, but you don't take it to cure cancer.'

Welcome, then, to the diet section.

A

- - - - - - - -

Alcohol

Alcohol is bad for us – but in moderation, for those without related health issues, the risk is sufficiently low that it's perfectly reasonable to enjoy a drink.

Most of the things we consume have pros and cons as far as a contribution to a healthy diet goes. For alcohol, though, the only thing to be said is that in moderation its risks are relatively low, so tolerable.

You may have seen newspaper articles saying that, for instance, red wine is good for you. This is a rather mixed assessment, as we'll explore in the **Red wine** section, but this is a result of the many other constituents of red wine. Alcohol itself, a simple organic compound, is a poison, pure and simple – but one that we can tolerate in low doses and that has sufficient pleasurable effects to make it worth tolerating.

Similarly, there was a lot of coverage in 2014 of a claim that anything up to a bottle of wine a day is fine. This came from a retired professor who did not present any evidence to back up his claim, when there is a huge amount of evidence for the harmful nature of drinking more than recommended amounts. In any science you will get mavericks coming up with an alternative view in good faith, but the only sensible approach is to go with the view held by the majority unless there is remarkable new evidence to suggest otherwise. No such evidence was provided here.

One myth is worth dismissing immediately. Alcohol is alcohol, and it doesn't matter what type of drink it is in. Mixing drinks makes no difference. There are some drinks, like whisky, that contain a range of other chemicals that are likely to make a hangover worse, but in terms of the impact on the body of alcohol itself, there is no difference. Some people think the mixing effect in cocktails somehow makes them more potent – it doesn't. But because many cocktails combine a high alcohol content with enough sweetness to conceal its potency, it can be easy to consume more alcohol than you realise when drinking cocktails.

The risks from moderate consumption – beyond a hangover – are usually due to inappropriate or illegal behaviour when we've had a couple too many, whether it is driving a car or simply doing things we wouldn't normally do and may regret afterwards. However, for heavier drinkers there are a number of concerns.

Most alcoholic drinks contain significant calories. Heavy beer drinkers, particularly, will tend to pile on the weight, as a pint of beer has as many calories as a packet of crisps, while the average wine drinker will consume around 2,000 kcal (see **Calorie intake**, page 17) a month from alcohol. This isn't a huge daily calorie consumption, but it is significant.

Alcohol is also a carcinogen – it causes cancer. In fact, alcohol is by far the biggest direct cancer-causing substance in our diets. It also increases blood pressure and risk of heart attack. And then there is the impact of the alcohol on the systems that take alcohol out of the body. The liver particularly can suffer with excess alcohol consumption, in the extreme case failing altogether.

Typical recommendations are that men do not regularly drink more than three or four units a day, and women do not regularly drink more than two or three units a day. We should know what units are by now, but they still cause confusion. Half a pint of 4 per cent alcohol beer is around one unit, a small glass of wine (125ml) is 1.5 units, and a single spirit is one unit. It is also recommended that you go at least two days a week without any alcohol.

To get a feel for the impact, if you go over the recommended limits to between five and eight units (men) or four and six units (women), you are 1.8–2.5 times (men) or 1.2–1.7 times as likely to get cancer of the mouth, neck and throat. Women are 1.2 times as likely to get breast cancer. Men are twice as likely to develop liver cirrhosis, and women are 1.7 times as likely. And men are 1.8 times as likely to develop high blood pressure, where women are 1.3 times as likely. Go beyond those limits into the higher risk zone and you can at least double those risks.

The best advice is still that pregnant women, or those trying to conceive, do not drink alcohol at all, and certainly don't exceed one to two units a week.

LINKS:
- **Calorie intake** – page 17
- **Hangovers** – page 291
- **Red wine** – page 87

Antioxidants

Antioxidants are vital chemicals used by the body to combat dangerous free radicals. But all the evidence is that consuming antioxidants has no benefit and could have some negative effects.

If you were to believe the advertising for some products, particularly those making use of the 'superfruits' tag that are rich in antioxidants, you would think that antioxidants were an ideal nutrient that provides huge benefits to make your body healthier. In fact, the picture is far less clear.

Antioxidants are naturally occurring chemicals that the body uses to counter the impact of free radicals, which are highly reactive substances that can damage DNA and cells, leading to cancer, cardiovascular problems and diabetes. Some free radicals play important roles in the body, but in the wrong place at the wrong time they are dangerous, and antioxidants are there to mop them up.

It seems reasonable, then, that tucking into products that are rich in antioxidants or taking antioxidant supplements would be a good thing. But it is often the case that just because something has an effect within the body does not mean that consuming it will have any direct impact. And even when it does, once you have enough of anything, adding more and more does not provide a benefit. At best the excess will be excreted and at worst it can have negative effects itself.

We get plenty of antioxidants from a normal diet, plus the antioxidants such as glutathione that the body manufactures itself. It might at first seem reasonable that 'if some antioxidants are good, lots of antioxidants are better. The more you take in, the better.' But think of applying that to eating in general. It's pretty obvious that 'if some food is good, lots of food is better. The more you eat, the better,' is wrong – and the same goes for antioxidants.

What is sometimes forgotten is that almost everything is damaging to the body or poisonous in excess. Toxicity is all about dosage. Water, for instance, does damage and can even kill if drunk to excessive levels. The antioxidant levels in foods – even superfruits – are sufficiently low that it would be difficult to overdose dangerously, and fruit has other benefits (though even fruit should not be taken in excess as it is high in sugar), but the real danger with antioxidants is when taken in supplements, where it is easy to exceed recommended daily amounts.

There is now good evidence that those taking antioxidant supplements on a regular basis are more likely to die prematurely than those who don't. (Specifically this seems to apply to vitamin A, vitamin E and beta-carotene supplements.) One reason for this seems to be that the supplements encourage growth in cancer cells, and so result in a greater likelihood of death for those already suffering from the disease. Another possibility is that increasing consumption of antioxidants means that our bodies' natural production of them tends to decrease – and those internally produced antioxidants have a much more significant impact. So, supplements could actually reduce our antioxidant defences.

The message, then, is that it is not a good idea to take antioxidant supplements, and while there is no harm in eating blueberries or cranberries or other fruits that are sources of antioxidants, they are unlikely to be giving any benefit. Just eat them to enjoy them!

LINKS:
- **Superfruits** – page 104
- Water excess – see **Hydration** – page 63

Artificial sweeteners

Artificial sweeteners are important in the fight to reduce sugar in our diets and have been shown to be safe.

Although aspartame (appearing under the brand name Nutrasweet) has been in use since the 1970s as a sugar substitute, there are still many who regard it with suspicion and claim that it is responsible for many health issues. A typical website I discovered researching this article claimed: 'Artificial sweeteners can actually be far worse for you than sugar and fructose, and scientific evidence backs up that conclusion.' This is just not true. It is important that we clarify this as, for health reasons, we are all being encouraged to consume fewer sugary products, and it can often be easier to switch to an artificial sweetener than to give up the sweet product altogether.

The European Food Safety Authority (EFSA) has come to a clear scientific consensus that aspartame is entirely safe. There are a number of reasons for this. First and foremost, numerous studies have shown that aspartame never makes it into the bloodstream. It is very quickly broken down to constituents that are found in almost all animal and plant protein.

Despite the claims of conspiracy theory websites, there is no scientific evidence of aspartame having any carcinogenic effect or of causing any genetic faults. Another frequent assertion is that consumption of aspartame during pregnancy results in an increase in asthma and allergic rhinitis in children; so this has specifically been studied, and once again there was no evidence. The link is fictional.

Like all substances, aspartame should not be consumed to excess. The potential harmful dose, however, would require you to consume several hundred cans of a diet drink a day. On a precautionary level, the EFSA suggests not consuming more than 40mg of aspartame per day for each kilogram of body weight. However, that still means that it's fine to drink more than is sensible. For a ten year old this is around seven cans a day, while adult women would be allowed fifteen cans and adult men nineteen. If you are drinking this many cans of fizzy drink, your diet is in need of a serious overhaul. Similarly, if you are using aspartame to sweeten your tea or coffee, the equivalent limits are around 36 spoonfuls/tablets per day for a ten year old, 77 for a woman and 98 for a man.

There are other artificial sweeteners. You will sometimes still see saccharin (brand name Sweet'N Low). A 1980s cancer scare pushed this out of favour, though the problem seems only to be with rats. Most consumers find the taste of aspartame more acceptable than saccharin. Some products use an extract of the stevia plant (brand names include Truvia and Rebiana), which is technically not an artificial sweetener, but a substitute, natural low-calorie sweetener instead. Stevia has a 'generally regarded as safe' status but has not had anywhere near the level of testing that aspartame has, and has produced some mixed lab results. So it may be worth treating with caution until more information is available, although it is used more and more widely. The most popular alternative is sucralose (brand name Splenda), which outsells aspartame in the UK. All studies have shown it to be safe, though it has had nowhere near as much scrutiny as aspartame. It has the advantage of not being sensitive to heat, so can be used in baking.

Whatever the sweetener, be wary of the 'seatbelts cause accidents' effect. There is reasonably good evidence that the more safety features a car has, the more careless our driving is likely to be, as we feel safer and so take more risks. Similarly, it can be tempting if you are 'being good' and cutting out sugar in your drinks to feel that you can now get away with more sugary or high-calorie treats. It's important to detach the two in your mind. If you can substitute, say, a sugar-free drink for the sugary equivalent but not change your diet otherwise, you are on the right track.

Research published in September 2014 suggested that artificial sweeteners could cause glucose intolerance in mice by altering the balance of gut bacteria. This could increase the risk of diabetes. As yet this is a single trial, and it is often the case that an effect in mice is not replicated in humans. This certainly does not make artificial sweeteners worse than sugar, and more research is needed – but it adds more weight to the ideal being to avoid all sweeteners if possible.

Artificial sweeteners are certainly not always the best solution. Rather than go from sugar to sweeteners in my coffee, I went to using no sweetener at all. For two weeks it tasted foul, but once I got used it, there was no problem, and now sweetened coffee tastes horribly over-sweet. There are plenty of circumstances where you can reduce or cut out sugar entirely. But in something like a can of cola the choice is really only sugar or sweetener, and it's worth going for the diet version if you can. (Again, it can take a few weeks to get used to the different taste, but for most it will grow on you.)

LINKS:
- **Sugar** – page 102

B

. . . .

Breakfast

**There are probably more sayings about breakfast
than any other meal – but do they make sense?**

Breakfast, we are told, is the most important meal of the day. But is this true? And if so, what should we be looking for in a breakfast? Is it best to go for a spartan muesli or the full English fry-up with all the trimmings?

Certainly, breakfast is a good thing. It sets you up for the day, making you more alert and productive, and research suggests that those who breakfast regularly tend to eat less (and snack less) later on. Those who regularly skip breakfast tend to have higher risk of heart problems. In part this is because extended fasting can result in increased blood pressure and raised cholesterol, and in part because of that tendency to substitute unhealthy snacks through the day. Though, as always with observational studies, it could also be that people who skip breakfast also tend to have more stress in their life, sleep less or have other factors that could have a negative effect on health. The lack of breakfast itself is not necessarily the cause.

In health terms, there seem to be benefits from having a relatively small number of meals – two or three – a day rather than snacking throughout the day. This seems to run counter to a frequently heard suggestion that frequent mini-meals are better for your digestion, but the problem with the grazing approach is that it is very difficult to keep on top of calorie intake when snacking, and all too easy to overdo intake of sugar, salt and fat.

It's best to avoid high-sugar foods for breakfast, as the result will typically be a dip in blood sugar a couple of hours later, just when the potential to go for a sugary snack is at its highest. It's particularly important to keep an eye on sugary children's cereals, which some of us continue to enjoy well into adulthood.

But what about the full English (or Scottish, Welsh or Irish) in all its fatty goodness? The positive news is that the recent discovery that saturated fat is not particularly worse than unsaturated fat for heart disease risks makes the traditional breakfast slightly less terrifying from a nutritional standpoint. However, it is still a very high-calorie, high-fat meal with a lot of processed meat. It came into being when workers were undertaking heavy manual labour and could easily burn up 5–6,000 calories a day, so could cope with the circa 1,500 calories provided by a typical fry up. But most of us don't work like that any more.

You may have seen in the newspapers that 'a fry-up for breakfast could be the healthiest start to the day'. Like most of these stories, this was based on a study that really doesn't tell us a lot about people. It was done on mice, which weren't, as you might expect, fed a full English, but rather high-fat mouse food. What the study does suggest is that it might be easier (for mice) to metabolise fat in the morning than later in the day, but that certainly isn't enough to suggest that a heavy fried meal makes the best breakfast.

Like pretty well any other dietary restriction, the fry-up is something that we shouldn't worry too much about enjoying occasionally, but if you have it more than once a week (one in four in the UK have it at least twice at a weekend), it would be a good idea to consider a change of diet.

LINKS:
- **Processed meats** – page 82
- **Saturated fat** – page 82
- **Sugar** – page 102

Burned food

**Although the risk is relatively low, there
is evidence that some over-cooked
foods can increase cancer risks.**

We've all burned the toast before, and have probably come up with
that old chestnut: 'It's charcoal – good for your digestion.' However,
very little of the blackened toast actually is charcoal, and the substance that is used medically to absorb poisons and quieten the
digestive system is *activated* charcoal, which has been treated to fill
it with little holes and bumps, vastly increasing its surface area to
enable it to act effectively. It is not just a lump of burned wood. In
fact burned food – or even just some of the over-cooked variety –
does seem to present us with a health risk.

When we grill, toast, roast or fry foods that are high in carbohydrates, we increase the quantity of acrylamides in them.
Acrylamides are relatively simple organic compounds, chemicals
that are poisonous in high doses and seem to present a cancer risk.
The more coloured a food is by the cooking, particularly in a high-
carbohydrate food, the more acrylamides it is likely to contain. So,
for instance, burned toast or deep brown, over-fried chips do have
an increased risk factor. A continued diet heavy in acrylamides
could double the risk of some cancers, though this may well only
be an increased risk of under half a per cent.

Another slightly surprising source is potatoes that have been
kept in the fridge. This can increase the amount of acrylamide
that forms when they are cooked (particularly if they are fried or
roasted). The best place to keep your potatoes is in a cool, dark
place – but out of the fridge.

There is also some acrylamide formation in coffee, but there's
not a lot we can do about this, as it occurs when the beans are
roasted. Again, the risk is very low – best to grin and bear it.

We can't totally avoid acrylamides. They occur in most cooked foods – and we get far greater benefits from cooking than we do problems. And eating the odd bit of burned toast or chargrilled potatoes will have a negligible impact. But by avoiding overdoing it, particularly avoiding food that has been fried, roasted or grilled until dark, you are certainly minimising the risk.

LINKS:
- **Coffee** – page 29

C

· · · · · · · ·

Calorie intake

**Calories measure the energy content of
food and can give us a good, quick measure
of just how much we are consuming.**

We are used to seeing food packaging that shows a measurement
of the calories it contains. Even some restaurants, notably the fast
food chains, now show the calorie content of their products on their
menus. This is a good thing, because keeping an eye on calories is
one of the easiest ways to avoid overeating and to keep your diet
healthy. It's not enough – you need to watch your sugar and salt
levels too, for instance – but it's a good starting point.

Note that calories and fat content aren't the same thing. Fat is a
major contributor to the calorie content of a food, but even a fat-free
product is likely to have calories from other food groups. A typical
fat-free yoghurt, for instance, has around 100 calories.

Another potentially confusing aspect of calories is the unit they
are measured in. You will see both 'calories' and 'kcal' or 'kilocalories' on labels. A kilocalorie is 1,000 calories. All dietary measurement is in kilocalories, but dieticians originally thought their
clients couldn't cope with that complicated 'kilo' bit, so shortened
kilocalorie to 'Calorie' with a capital C. So a Calorie is 1,000 calories. This is very confusing, particularly as 'Calorie' is often written without the capital letter. Increasingly the standard is to have
all labelling as kilocalories (kcal), and that's what we'll use here.
Worse still, the calorie is an outdated unit – it was replaced in the
scientific world by the joule a good 50 years ago, but the dietary
world is slow to catch up.

Calories provide a measure of energy. They show how much

energy content there is in a food. There is then a simple balance. If you burn off more energy through exercise than your body absorbs from your food (not all the calories you eat will be retained – some will pass through), you will lose weight. If your body takes up more calories than you burn, you will gain weight.

The rule of thumb is that an average man needs around 2,500 kcal a day and an average woman around 2,000 kcal. If you don't do much exercise, reduce those to 2,000 and 1,700. This is where those menu boards in fast food restaurants can be quite scary. Go for KFC's 'Big Daddy box meal' with regular Pepsi, for instance, and you are consuming 1,615 kcal in a single meal. Even for a man undergoing a reasonable level of exercise that's 65 per cent of his daily calories gone in one go. And British fast food meals are skimpy when compared with American portions.

That's really all you need to know to diet successfully. Reduce calorie intake and increase exercise and you will lose weight. The difficulty is managing this consistently, as most dieters follow the diet for a relatively short period of time, then binge and, if anything, put on weight. It is also very difficult to lose weight by exercise alone. Time and again, studies have shown surprisingly low weight loss when overweight people tried to reduce weight with an exercise regime. This seems to be because they compensated for the extra work they were doing by eating more and by moving less at other times than the specific exercise period.

What seems to be one of the most reliable and easiest diets to stick to is the 5:2 diet, where two days a week you eat a significantly reduced calorie intake, while the rest of the week you eat normally. (This is sometimes described as 'whatever you like', but the idea is not that you can pig out on piles of fatty food five days a week, just that there is no specific food restricted.) The reduced calorie days should be between half and one third of the usual daily intake. This sounds like a tiny amount, but it's easily done with a light breakfast, a sandwich for lunch and a low-calorie meal for dinner. What's important is to keep an eye on the extras like drinks and snacks that can easily push you over the top.

LINKS:
- **Fat** – page 44

Carbohydrates

Fads come and go as to whether carbohydrates are a great way to fill up without fat or are to be avoided at all costs. But what's the reality?

A carbohydrate is a relatively simple organic compound, made up of just carbon, oxygen and hydrogen, with the hydrogen and oxygen usually in the same 2:1 ratio as water. Technically sugars are carbohydrates, as is the fibre that is so good for our digestion, but the term is usually used in food to refer to the long-chain or polymer versions of carbohydrates, sometimes called polysaccharides, of which a typical example is starch. Such molecules are energy stores.

In the stuff we eat, most of the carbohydrates come from the 'staples' like potatoes, bread, rice, pasta and cereals. It is generally recommended that for a balanced diet, around one third of the food we eat should be carbohydrates. Often there are different versions of the carbohydrate foods, some of which are better nutritionally – usually where they contain extra fibre. So potatoes are better skin-on, and the 'brown', wholegrain versions of most other carbohydrate sources tend to be healthier.

Going on dietary surveys, a fair proportion of the population don't have enough starchy foods in their diets. For some this is because they have adopted a low-carb diet – but there is no good evidence that these are beneficial, and they can result in poor nutrition. In fact, cutting carbs isn't a great way to reduce calories as carbohydrates contain less than half the calories of the same weight of fat. Studies suggest that a healthy, balanced diet gives just as much weight loss as low-carb diets, but typically has a better fibre and mineral content.

It's not a great thing to miss out on starchy foods, as they combine a good source of energy with the main source we have of fibre, calcium, iron and B vitamins. Don't, for instance, dismiss the humble potato. We tend to look down on it because of its familiarity, but it gives us energy, fibre (if skin-on), potassium, B vitamins and a fair amount of vitamin C.

When cooking starchy foods, try not to overcook them when toasting, roasting, baking, grilling or frying, as this can result in an increase in levels of acrylamides, which are poisonous in large quantities and may carry a cancer risk. It's also worth storing your potatoes outside the fridge (ideally in a cool, dark place), as low temperatures can result in sugar forming, which then encourages acrylamides to form. Cut off green parts of potatoes too – these tasty tubers are of the nightshade family, and the green parts have the family tendency to be poisonous.

LINKS:
- 5:2 diet – see **Calorie intake** – page 17
- Acrylamides – see **Burned food** – page 15
- **Fat** – page 44
- **Starch** – page 99

Chemicals

**All foods are made up of chemicals,
so beware of claims that a product is
'chemical free', and don't be put off
by a scary looking contents list.**

The contents list that manufacturers are required to put on many foods can look scary – and there are things to keep an eye out for – but the mere presence of a long string of E-numbers and other nasty sounding chemicals is not in itself a problem.

Let's take a specific content list and see what we should look out for:

Aqua 84%, sugars 10% (of which fructose 48%, glucose 40%, sucrose 2%), fibre 2.4% (E460, E461, E462, E464, E466, E467), amino acids (glutamic acid 23%, aspartic acid 18%, leucine 17%, arginine 8%, alanine 4%, valine 4%, glycine 4%, proline 4%, isoleucine 4%, serine 4%, threonine 3%, phenylalanine 2%, lysine 2%, methionine 2%, tyrosine 1%, histidine 1%, cysteine 1%, tryptophan <1%), fatty acids <1% (linoleic acid 30%, linolenic acid 19%, oleic acid 18%, palmitic acid 6%, stearic acid 2%, palmitoleic acid <1%), ash <1%, phytosterols, oxalic acid, E300, E306, thiamine, colours (E163a, E163b, E163e, E163f, E160), flavours (ethyl ethanoate, 4-methyl butyraldehyde, 2-methyl butyraldehyde, pentanal, methylbutyrate, octene, hexanal, styrene, nonane, non-1-ene, linalool, citral, benzaldehyde, butylated hydroxytoluene (E321), methylparaben, E1510, E300, E440, E421, aeris (E941, E948, E290)

It's a scary looking list, and any concerned parent or anyone worried about E-numbers would be inclined to steer well clear. The good news is that by far the biggest content is water (usually called 'aqua' in contents lists, particularly in cosmetics, so it sounds more impressive). Unfortunately, though, a 10 per cent sugar content is not great, there are both poisons and carcinogens in there, and there are enough E-numbers to imagine it would send the average hyperactive child into overdrive.

The interesting thing, though, is that this is not a highly-processed food; it is the contents list of that most revered of superfruits, the blueberry. Not some sugary blueberry concoction, but the actual fruit itself. We don't usually see the chemical contents lists of fruit and veg, but it would make a useful comparison with the products that really worry us if we did.

The level of sugar in fruit is fine as long we don't overdose on it. As for the poisons, pretty well every food has poisons in it. Most plants, for instance, contain a range of natural pesticides that are deadly in large enough proportions, but, as is always the case with poisons, it is the quantity that matters, and here the volumes are far too low to have any significant impact.

As for those E-numbers, it is a sobering reminder that pretty well everything natural has an E-number, and the presence of E-numbers in a food is not a problem per se. E-numbers are simply a European Union labelling standard for all food additives; they are not intended to say whether those additives are good, bad or indifferent.

So let's be clear. Whether we are talking about diet, health or any other products, there is no such thing as a chemical-free substance or object. Light is chemical free, but that's about all you can get. If someone claims a product is chemical free, they are either ignorant or lying. And there is no way you can avoid chemicals by having a particular diet or buying particular products. There is no difference whatsoever between a synthetic and a natural chemical – how can there be, when a chemical is just a compound of elements? It doesn't know where it came from. But, often, natural sources of chemicals are more dangerous, as they are far more likely to be impure, with all kinds of nasty substances coming along for the ride.

What's more, we are far safer now than our ancestors were, as we have greater protection against the misuse of chemicals. In the old days, people happily plastered toxic lead oxide on their skin to whiten it, and used wallpaper coloured green by deadly arsenic, which gave off poisonous fumes. Though things go wrong sometimes, the picture that 'back to nature' activists paint of a world where we are increasingly subject to dire chemicals gets reality back to front.

Greenpeace comments on the 'Chemicals out of control' section of its international website:

> If someone came into your house, mixed you a cocktail of unknown chemicals – and offered you a drink – would you take it? Of course not. You wouldn't want untested chemicals in your home, your drink, or your body. You don't want them – but shockingly – they're already there.

But actually most of the cocktails of unknown and untested chemicals we come across are in natural substances. A new synthetic product has to be extensively tested, but only a tiny handful of the 'chemical cocktails' that make up tea and coffee have ever been tested (and many of those have proved to be dangerous in large quantities). That Greenpeace comment is a triumph of hype over reality.

A small confession – I made up the name of the final ingredient in the blueberry to give it a name as exotic-sounding as 'aqua' is for water. 'Aeris' is just common or garden air.

Chlorophyll

Chlorophyll (as opposed to Cholesterol – see page 28) is the stuff that makes plants green and is used in photosynthesis. Some nutritionists and nutrition websites recommend chlorophyll in a diet, but there are no proven health benefits.

Chlorophyll, the green substance in plants, does a superb job in photosynthesis, where plants gain energy from sunlight and absorb carbon dioxide to build their structure, along the way releasing the oxygen that enables animals like us to breathe. It is sometimes suggested that this means that chlorophyll is also good as part of a human diet, because it somehow enables us to take in the energy of the sun or because it will oxygenate the blood.

In practice, there is no evidence whatsoever of this happening. It would be very strange if it did. Chlorophyll's role is to enable photosynthesis, which requires light – not something food gets a lot of once you swallow it. Chlorophyll doesn't *carry* oxygen, the photosynthesis process produces oxygen as a by-product. Even if the chlorophyll were churning out oxygen in your stomach (which wouldn't be safe as your stomach contains flammable gas and adding oxygen would make it like the reaction chamber of a rocket), there is no mechanism for the oxygen to get from there to your bloodstream.

As for harnessing the energy of the sun, as if eating our greens turned us into plants, it's hard to imagine any way this could be possible. Websites often hint that chlorophyll is similar to the haemoglobin that carries oxygen in our blood (and it is). But being similar isn't good enough in body chemistry. What certainly isn't true, as I read on one website, is that: 'According to scientist [*sic*], the body has the ability to convert chlorophyll to haemoglobin by changing just one little molecule [*sic*] of magnesium into iron.' No, I'm sorry, according to *scientist*, the body has no such ability.

By all means, eat green foodstuffs from spinach to broccoli, which are all packed with plenty of essential nutrients. But their chlorophyll content is not one of those benefits.

Chocolate

For most of us, chocolate is a wonderful treat, and, as long as it's consumed in moderation, it's not a bad thing, either.

Chocolatiers will tell you that the reason chocolate is so appealing is because of the tactile sensation of consuming it. It melts at the temperature of your mouth, so the solid turns into a sensuous liquid on your tongue. There is an element of truth in this. It is also the case that there are a range of active chemicals in chocolate that influence our brains, and sugar is certainly part of the attraction of the modern version of chocolate. But there seems little doubt that we also get a kick from a substance called theobromine.

This is a bitter tasting alkaloid, a term we often associate with drugs like morphine and natural poisons. Caffeine, nicotine, quinine and cocaine are all part of the alkaloid family, but none has the appeal of theobromine. A clue might come from the Greek meaning of the name, which is roughly 'food of the gods'. Theobromine is the compound that makes chocolate special and is found in the cocoa tree. The seeds of this tree (misleadingly called cocoa *beans*) contain the fatty substance cocoa butter that is the main ingredient of chocolate.

Chocolate has been enjoyed as a drink in Central and South America for at least 3,000 years, and has been popular in Europe since the 17th century. In its original form, the drink was bitter (it often had chillies added to give it extra bite) – it was a European twist to add sugar and milk to make something closer to modern drinking chocolate. The familiar solid form didn't arrive until the 19th century, which was also when theobromine was discovered. The substance has similar effects on the brain to caffeine, which is probably why you will occasionally see it said (incorrectly) that chocolate contains caffeine. Theobromine can reduce sleepiness and in large quantities produces a jittery sensation. On the positive side, it is a cough suppressant and can help reduce asthma symptoms.

Most of us have heard that chocolate is bad for dogs – it is theobromine that is to blame. The darker the chocolate, the higher the concentration of theobromine, and the more dangerous it becomes. A small dog could be killed by as little as 50g of strong dark chocolate. Smaller doses will cause vomiting. This isn't a problem limited to dogs; poisoning occurs in all mammals to some degree, though the speed at which theobromine is disposed of by the system differs from species to species. Cats are particularly sensitive to theobromine but rarely eat chocolate because they don't have sweet taste receptors and so don't get the kick from sweets that humans (and dogs) do.

Theobromine is also poisonous to humans, though not to the same the degree as it is to dogs, and shouldn't cause concern. Almost everything is poisonous in a large enough dose (even water, for example), and toxicity is all about dosage. In the case of theobromine, humans have about three times the resistance per kilogram of bodyweight as does a dog and are significantly heavier, so we are much less likely to be damaged by our treats. A dangerous dose for an adult human would involve eating more than 5kg of milk chocolate.

A number of health benefits have been claimed for chocolate, including reducing blood pressure, reducing stress, reducing diabetes risk and giving limited protection against bowel cancer. None of the trials that have come up with these results have been big enough or repeated sufficiently to be sure of the outcome. The effect on blood pressure was slight, the cancer results are only laboratory-based, and the stress test was poorly designed (and sponsored by a chocolate manufacturer).

The diabetes results were based on compounds in chocolate called flavonoids, but the trial could not show if the flavonoids caused the benefits – and chocolate is not the best source of flavonoids anyway (the study focused mainly on berries and wine). And, of course, we know that excess sugar consumption, a major content of most chocolate products, makes diabetes more likely.

Bearing in mind the high fat levels in chocolate, the balance of evidence is that we can't think of it as a healthy food, but one to enjoy in moderation.

LINKS:
- Caffeine – see **Coffee** – page 29
- **Fat** – page 44
- **Sugar** – page 102

Cholesterol

There is a lot of confusion about cholesterol.
It is certainly far too simplistic to say that
cholesterol is bad for us, but raised levels
of some types of cholesterol are strong
indicators of risk of heart attack and stroke.

The good news about cholesterol is that it is an essential chemical in the body, required for the cell membranes – the outer layers that hold the contents of the cells in our body in place – to allow various chemicals in and out of these complex tiny factories. Broadly, cholesterol comes in two forms, the larger low-density (LDL) form and the smaller high-density (HDL) form. LDL is sometimes called 'bad cholesterol' as it can transport fat into artery walls, while HDL is nicknamed 'good cholesterol' as it can remove some fat molecules from special cells in the artery walls.

For a long time it was thought that cholesterol from foods like eggs was bad for us, but it has been shown that this doesn't tend to increase levels of cholesterol in the blood, which is where the danger from cholesterol lies. The worst dietary culprit seems to be trans fat, which increases LDL and lowers HDL levels in the blood. Cholesterol levels are also increased by smoking or having diabetes or high blood pressure, and are often found to be naturally high in those with a family history of stroke or heart disease.

A good, healthy, low-fat diet can contribute to keeping LDL levels down, as can regular exercise, but some of us are naturally more prone than others to high LDL levels and the associated risk. Levels can be reduced to a small degree by plant sterols, which are added to some brands of yoghurt and spread marketed as cholesterol reducers, though strangely there seems little evidence of a beneficial effect on cardiovascular disease.

There is stronger evidence for the ability of the drug family known as statins to reduce both cholesterol levels and heart disease risk. These are now widely prescribed for those with the potential for such problems. There has been something of a spat between medics supporting the wide use of statins and other clinicians who feel that the side effects of statins (most commonly stomach problems) mean that they should not be too widely prescribed, but both sides agree that they should be taken by those with heightened LDL levels and family history of heart disease. It is only the preventative use of statins by those with no risk factors that has been called into question. If in doubt, go with your GP's guidance.

LINKS:
- **Eggs** – page 41
- **Fat** – page 44
- **Trans fats** – page 109

Coffee

Once you start to look at the science of coffee, things can start to seem more than a little worrying. But there's good news too.

For many of us, coffee is the indispensable kick-start to the day. And it's a huge help to get through those mid-morning and mid-afternoon dips. Not to mention an essential after a good meal. But there is no doubt that coffee gives us one or two things to worry about.

A single cup of coffee contains the same quantity of cancer-causing chemicals (carcinogens) as a whole years' consumption of agrochemical residues in our diet. And that's just what we know about. Coffee is a very complex mix of chemicals – around 1,000 in all. So far only 30 of these have had the kind of high dosage testing usually done on dangerous chemicals, and 21 proved carcinogenic. However, it ought to be stressed that the risks involved are still extremely low – in practice it takes over six cups a day to have any measurable risk, and even then it is small. (A cup here is around 250ml, not a massive, top-of-the range coffee-shop bucket, which will be the equivalent of at least three cups.)

For some people, coffee (caffeine in particular) is a problem when it comes to sleep, and a study in 2013 did show that caffeine could disrupt sleep for up to six hours after drinking it. However, there has to be a little caution with these results. It was a very small study with only twelve participants completing it, and it has yet to be replicated. All the evidence is that responses to caffeine are very variable. So, yes, if you are significantly influenced by caffeine it would be best to stay off the coffee for several hours before sleeping – but just as many people seem not to be affected. If in doubt, go for a low-caffeine option like an instant coffee or a decaf version.

Caffeine is also a diuretic, which encourages you to pass urine. This can produce some dehydration if taken to excess, but there is reasonable evidence that drinking two or three cups has very little dehydrating effect on a healthy person. Although there isn't strong evidence on danger from the impact of drinking coffee when pregnant, it does appear that caffeine will pass through the placenta and get to the foetus, which is very sensitive to the drug. That being the case, it makes sense to minimise caffeine consumption when pregnant.

One other slightly surprising thing is that filter coffee may be better for you than cafetière-made coffee or the espresso version usual in a coffee shop. This is because one of the other chemical constituents of coffee, cafestol, pushes up the 'bad' LDL cholesterol level. Cafestol is found in the oily part of coffee, which gets soaked up in a filter and doesn't end up in your drink. By comparison, cafetières and espresso machines don't have anything to filter out the oily part, so it reaches your cup. (More cafestol gets through from the cafetière, perhaps because the coffee isn't compacted, as it is in an espresso machine, which acts as a partial filter). If you have high cholesterol or are trying to keep it down, stick to filter coffee or instant.

On the plus side, studies have shown that coffee (more so than caffeine alone) does what we all think it does – it really does give a bit of a mental edge. There have also been some studies that suggest regular coffee consumption could protect to a small degree against type 2 diabetes, Parkinson's and some liver and heart problems – but as is often the case with this kind of research, it is hard to be sure that it was the coffee that really made the difference, rather than other lifestyle variations.

All the evidence is that, despite a tendency mentally to link coffee to much more dangerous consumption like alcohol and tobacco, it is in fact harmless for the vast majority of people, and considering the enjoyment it gives to many, it is not something most of us need to avoid. Of course, if you drink it to excess it will cause problems. High doses of caffeine can lead to jitteriness or tremors and feeling stressed and uncomfortable. And you do have to watch the fancy coffee-shop drinks that have syrup and whipped cream thrown in – these can weigh in with as many as 500 calories, around a quarter of your entire daily intake. But moderate coffee drinking is no problem.

LINKS:

- **Brain food** – page 158
- **Cholesterol** – page 28

D

. . . .

Dairy

**There have been few foods that have
yo-yoed more in dietary advice than dairy.
So what should we be doing about milk,
cream, butter, yoghurt and cheese?**

Although eggs somehow often end up under dairy, we're sticking here to the true products of the dairy – see the separate entry on **Eggs** (page 41). When I was young all we heard about dairy was the good stuff: how it built strong bones and was a great all-round food. But then the fear of fat came upon us, and dairy became something to avoid. Now it's back in a curiously uncertain position, needing a spot of clarification.

One thing that is worth mentioning is lactose intolerance. Milk was only ever intended in nature as a food for young animals, and it is normal for animals to stop being able to process it as food when no longer juvenile. The same originally went for humans, but most of us are mutants, capable of processing dairy all our lives. Some, though, still can't – a genetic variation particularly common in East Asia. Lactose intolerance (which is rare in Europeans) tends to show when symptoms like diarrhoea, stomach cramps and a feeling of bloating develop a few hours after consuming dairy. If you have these symptoms, consult your GP.

There are three interesting components in dairy – fat, calcium and sugar – and it's worth thinking briefly about each to get a balanced picture. Dairy products do indeed contain fat, though the percentage is not always what you might imagine. Full-fat milk, for instance, is typically 3–4 per cent fat, which is not exactly high. Personally, I use skimmed milk, which has practically no fat, because once you get used to it, there really is no disadvantage. However, given the percentage, milk is not our worst source of fat.

Yoghurt can be pretty much fat-free, though it varies a lot, but the two fat monsters are butter and cheese. Like it or not, butter is at least 80 per cent fat, and cheese typically ranges from 25–40 per cent (much of the rest being water). However, this isn't as bad as it sounds. Bear in mind that the alternatives to butter, like olive oil, are pretty much all fat too. Admittedly, you can get lower-fat spreads, which primarily use water to decrease the fat content (so if you fry with them, you get just as much fat in the usable frying material), but apart from the practicalities of spreading it when it's hard from the fridge, it's hard to beat butter as a spread for bread.

We used to frown on butter because it has more saturated fat than many spreads, but this now seems not to make any significant difference to any health problems compared with unsaturated fats, and we know that a certain amount of fat in the diet is a good thing. However, while it certainly is possible to have too much fat, and moderation is to be recommended, there's no reason not to enjoy butter and cheese.

When it comes to calcium, for once the old stories we were told are true. Dairy products *do* provide the calcium we need for healthy teeth and bones – not to mention to keep muscles, including the heart, functioning properly and to ensure that blood clots effectively. We need about 700mg of calcium a day as adults, and most of us get this happily from our dairy plus other good sources like leafy vegetables (except spinach) and nuts. It is possible to overdo calcium, so best not to take supplements unless they are prescribed. A 200ml glass of milk contains around 240mg of calcium, butter typically has around 40mg per 100g, yoghurt around 250mg per pot and cheese around 250mg in 40g.

The joker in the pack is sugar, which is not something we associate with dairy as it doesn't taste particularly sweet (masked, in part, by the fat). But a 200ml glass of milk contains around 10g of sugar, where the guideline daily amount is currently 70g for men and 50g for women, and the best evidence suggests it should be half that. So that's up to 20 per cent of your daily sugar intake in a single glass of milk. A 'no added sugar' yoghurt will typically come in at around 7g, compared to just 1g in 40g of cheese and a tiny amount in butter.

So, though low-fat milk might seem an ideal drink, it's best not to have more than two glasses a day, bearing in mind that will use up nearly half your adult sugar intake.

Overall, dairy is on the up again. As long as we bear in mind that sugar content for milk and yoghurt, and don't go mad on fat, it is a great contributor to a balanced diet.

LINKS:
- **Allergies** – page 254
- **Eggs** – page 41
- **Fat** – page 44
- **Spreads** – page 97

Detox

A whole range of products are on offer to help us 'detox', and after traditional times of over-consumption like Christmas and New Year, magazines push the detox message. But in reality there's no such thing.

There's nothing a woman's magazine likes better than the latest 'detox diet', and many health shops sell detox products. Even the Prince of Wales' Duchy Organics used to produce a 'Detox Tincture'. It's a perfectly reasonable sounding concept. There are lots of toxins (poisons) in the world. In fact, practically every plant we eat contains toxins that act as natural pesticides. But most of the poisons are in far too small a quantity to do us any harm, and inside our bodies the liver, kidneys and the whole digestive system is constantly busy removing toxins from our systems. So, what do detox products claim to do, and what do they actually do?

The picture the advertising gives us is that our bodies build up all sorts of nasty stuff (particularly after periods of heavy eating: January is peak 'detox' advertising time), and the only way to get rid of this is to eat or drink some wonder substance that will somehow flush it out. It's a bit like the chemicals put in car radiators or heating systems to try to move the gunge that builds up. However, we aren't radiators, and there is no evidence that the body needs this nor that these detox diets deliver any value. To quote *Bad Science* expert, Dr Ben Goldacre: 'The detox phenomenon is interesting because it represents one of the most grandiose innovations of marketers, lifestyle gurus and alternative therapists: the invention of a whole new physiological process.'

Of course, eating and drinking large quantities of stuff that's bad for us and partying all night will leave us in need of sleep, build up fat, clog the arteries and so forth. This will be improved by leaving off the bad things and behaving sensibly. But there is no identifiable 'detox' process happening, like flushing gunge out of a radiator. It is just that your body is recovering from the misuse perfectly naturally, with no magic external detox mechanism at work. As the toxicologist John Hoskins points out: 'The only thing that loses weight on a detox is your wallet.'

But surely even medical types recommend drinking water as a 'detox' technique after drinking alcohol? And why not boost the effect with a detox product? Certainly drinking water is an important thing to do, as alcohol dehydrates, among other nasty effects. Sir Colin Berry, a distinguished pathologist says:

One of the most poisonous chemicals that many people encounter is alcohol. However, even if you drink an almost lethal dose of alcohol (which I don't recommend) your liver will clear it in 36 hours without any assistance from detox tablets. As a pathologist, I am frustrated by the claims that a detox diet will somehow improve your liver function, the only thing you can do to help your liver after a period of indulgence is to stop drinking alcohol and drink water.

One more legitimate concern is that there are some unpleasant substances that build up in the body over time. They tend to be nasties like PCBs and dioxins, used in some manufacturing processes, and some pesticides, which dissolve in fat and can accumulate in fatty tissue. To try to deal with this, the detox approach tends either to cut out solids altogether for a while or, at the very least, to greatly reduce solid intake.

Unfortunately, not only is this a very slow process – you would have to fast for years to get rid of a typical build-up – the main impact of the detox attempt is to move the contaminants from fat, where they are harmless in small quantities, to the blood, where they can do more damage. The detox doesn't so much flush out the contaminants as make them more active.

Overall, then, the concept of 'detoxing' is a fictional one. The better message is simply to have a good, balanced diet with limited alcohol intake and plenty of exercise. It's boring – and it doesn't sell detox tinctures – but it's the best you can do.

LINKS:
- **Alcohol** – page 5

Diets

**There really is no magic secret to dieting –
but some diets are better than others.**

I have already given away the secret to a healthy diet in the introduction to this section, but in case you are the kind of person who doesn't read introductions, there really is no need for all those fancy diet books. This is all the diet guidance you need:

Don't eat too much – if you are putting on weight, eat less. Eat plenty of fruit and vegetables (and don't make them into smoothies as this ruins the valuable fibre), including around one third starchy foods, preferably wholegrain. Don't eat too much processed meat. Drink alcohol moderately, if you must. Avoid sugar and salt as much as possible and don't go overboard on fats, avoiding trans fats entirely. It's not strictly part of a diet, but add 'don't smoke' and 'take sensible exercise' and you've got an instant health plan.

Many of us have a go at a diet when we feel we need to make a change – say after Christmas and New Year. But all the evidence is that short-term diets have no long-lasting effect. You may well lose a kilo or two for a few weeks, but once you quit the diet – and if it's a fancy diet you certainly will – the weight is likely to return. In fact, a lot of people end up slightly heavier once they have gone through the 'recovery' overeating phase following a diet.

There are one or two diets worthy of specific comment. The one that makes most sense is the 5:2 diet (see **Calorie intake**, page 17). What we are looking for in an effective diet that will make a sustained change is something that will reduce calorie intake in a way that you can maintain, and unlike all the weird and wonderful plans you see on the market, the 5:2 is generally easy to keep up, though most people will probably gradually loosen the limitation on the '2' low calorie days per week.

Diets that cause the most controversy are the low-carbohydrate/high-fat diets like the Atkins and Dukan diets. All the evidence is that we ought to significantly reduce sugar intake, and because the body converts carbohydrates to sugars, it is certainly good not to over-indulge on carbs. The theory of the Atkins diet is that the lack of carbohydrates forces the body to consume body fat instead, but there seems limited evidence that this is true, nor does it seem to be the case, as Atkins claimed, that consuming fat uses more food calories per calorie of work than consuming carbohydrates. There are serious problems with the way the Dukan diet restricts vegetables, as this makes no nutritional sense: the diet may be a good way to lose weight quickly, but isn't nutritionally balanced.

There are definitely other issues with low-carbohydrate diets as they typically impose no calorie restrictions, nor do they seem to consider the impact of salt. However, as sugars are usually included in carbohydrate counts on foods, the diet should at least keep your sugar intake down. When undertaking a low-carbohydrate diet, it is important not to let processed meat consumption (for instance, bacon and sausages) go up as part of the attempt to boost protein intake. Overall, low-carbohydrate diets are unlikely to give any benefit over a traditional balanced diet, and there is one recent large study, the best yet undertaken on a low-carb/high-protein diet, that suggests that these diets increase the risk of strokes and heart problems, though more research is needed.

As for the rest of the swathe of diets you could try, there is simply no point doing anything faddy and weird, like trying to live entirely on grapefruits and carrots, because such diets don't make any nutritional sense, and are the kind you will definitely give up after a few weeks, bouncing back and overeating. So, why waste your money on the latest diet book? Do something fun with your cash instead.

LINKS:

E

• • • • • • • •

E-numbers

'E-numbers' are so closely associated with a bad diet and artificial additives that we need to remember that these are just an EU additive identification system and don't indicate if a product is good or bad.

In the 1960s, the predecessor of the EU put together a list of codes for food additives, covering substances like colours, preservatives, antioxidants, thickeners, stabilisers, emulsifiers, acidity regulators, sweeteners and all the bits and pieces that food manufacturers like to add to make sure the food we buy stays safe, good looking and tasty.

All these substances are chemicals – but it should be emphasised that this isn't a criticism. Every food is made up entirely of chemicals. If you ever see any product advertised as being 'free from chemicals', you should be buying a totally empty container, as even fresh air contains a range of chemical elements.

The E-number system includes both natural and artificially created chemicals. There is absolutely no distinction between the artificial and natural versions of any particular chemical compound, but the concerns often raised about E-numbers tend to focus on the synthesised compounds rather than, for instance, saffron, beta-carotene, vitamin C, citric acid or many more natural sounding substances, which all have E-numbers.

The E-number system also omits some of the worst food additives, such as salt, fat and sugar – all added to processed foods to make them more attractive to eat and all 'natural' but bad for us in excess. Of course, some E-numbered products are better for us than others, but don't be worried if a product contains a fair number of E-numbers. Bear in mind that the analysis of a blueberry – the fruit, with no additives – shows it to contain at least 21 E-numbers (see **Chemicals** for details of this). It's only because they're not on the label that they aren't a worry to us.

LINKS:
- **Chemicals** – page 21
- Hyperactivity – see **Hyperactivity and sugar** – page 299
- **Salt** – page 90

Eggs

**Not long ago we were discouraged from eating too
many eggs because of their cholesterol content
– but this has now been roundly dismissed, and
eggs form an excellent part of a balanced diet.**

A while ago, the advice was that we should eat no more than two eggs a week. The reason for this was that eggs contain a significant amount of cholesterol, and, as we all know, high cholesterol levels in the blood are indicators of increased risk of heart attack and stroke. What was not appreciated at the time – though it really should have been – is that consuming cholesterol does not mean that levels will increase in the blood.

This is similar to the way we now know that consuming lots of antioxidants does not give us significantly greater antioxidant levels internally. When you think about it, it makes a lot of sense. Just because a particular chemical is added to the highly active acid mix in your stomach does not mean that the chemical will make its way into your bloodstream unchanged by the digestive processes.

All the evidence now shows that we do not need to worry about cholesterol levels in eggs, which are a good source of protein and contain a range of vitamins and minerals, including vitamins A, B2 and D and iodine. Unlike cholesterol, our bodies have mechanisms for incorporating these nutrients into our systems. Of course, this doesn't mean we should eat a plateful of fried eggs a day. Boiled, scrambled and poached eggs are all lower in fat than fried eggs. We do need some fat – so a weekly fried egg or two will do no harm – it's just not something we want to overdo.

If there are health problems from eggs, it is more likely to be due to bacterial contamination, notably salmonella, which has the potential to cause food poisoning. Uncooked or lightly cooked eggs are best avoided by young children, the elderly and pregnant women. Commercial products like mayonnaise made with uncooked eggs use safe pasteurised eggs, but if you are in an at-risk group it's best to avoid homemade equivalents. As part of a balanced diet there is no problem with having an egg or two every day if you fancy it – an egg is a great foodstuff.

LINKS:
- **Antioxidants** – page 7
- **Cholesterol** – page 28

Enzyme powders

Enzymes help our stomachs break down complex structures, making them easier to digest. So it seems reasonable that adding extra enzymes would help improve digestion. But does it?

You don't have to look far in health food shops and dietary websites to find enzyme powders. The suggestion is made in selling them that our natural diets of raw foods contain enzymes that naturally help break down the food, but our modern processed diet is lacking in these natural digestives, so we can help things along by adding an enzyme powder to our diet.

The enzyme powders usually contain vegetable-based compounds, which supposedly boost our natural enzyme store. Unfortunately, this is a classic example of a common misconception. Adding to your diet something that it is useful to have in your body does not necessarily provide any advantage. We already have appropriate enzymes in our system. The 'alien' plant enzymes will just be digested like any other protein. In effect, an enzyme powder is just another form of protein shake.

LINKS:
- **Protein** – page 84
- **Supplements** – page 106

F

· · · · · · · · ·

Fat

**When it comes to diet, fat has a bad press.
But as a term it covers a whole mass of
substances, some of which are better for us
than others. And what we thought we always
knew turns out to be not quite right.**

Fats are organic compounds in the scientific sense, meaning that
they contain carbon. Fats combine glycerol (also known as glycerine) and fatty acids, which are chemicals with a long chain of carbon
atoms with a 'carboxyl' chunk on the end to make them acids. We
have tended to obsess about levels of fat in the diet, to the extent
that we ignored other substances (notably sugar and salt) that we
ought to be controlling with more rigour.

There is no doubt that too much fat is bad for us, but equally we
shouldn't avoid it entirely. We do need some fat in our diet. This was
discovered the hard way by early Canadian trappers, who lived on
a diet composed almost entirely of rabbits, which are so low in fat
that the trappers suffered from malnutrition, and some even died.
Specifically, there is very little evidence that fat causes us to become
obese – this seems primarily down to excessive carbohydrate consumption, particularly in the form of sugar.

The main thing with fat is to keep consumption to a reasonable
level, and to be aware of the different types of fat and their relative impact. The worst by far are trans fats, found in hydrogenated
fats, which increase the risk of heart disease and obesity. These are
often found in processed foods, cakes and biscuits. Then there are
saturated fats, typically of fatty meats and dairy. These used to be
on the 'avoid like the plague' list because they were also thought to

influence heart disease. The latest information, though, is that they do not increase risk. However, like all fats, excess consumption can contribute to health problems. Finally, there are the unsaturated fats, particularly polyunsaturated, which are often claimed to aid health, though there is increasing doubt about these benefits – and it is still important not to consume too much. See the separate entries on these kinds of fat for more detail.

There is no dietary difference between a solid fat, such as butter or lard, and a liquid fat like oil – the difference merely reflects the natural state of the substance at room temperature.

LINKS:
- **Saturated fat** – page 92
- **Sugar** – page 102
- **Trans fats** – page 109

Fibre

Dietary fibre is a strangely contradictory food. It is difficult to digest, so has limited food value, but is very valuable in keeping the gastrointestinal system healthy.

Fibre (which used to be called roughage) is made up of molecules in long chains, including cellulose – compounds not unlike a natural form of plastic. (In fact, cellulose was the basis of one of the first plastics, celluloid.) Most of us don't get enough fibre, with the average UK daily diet containing around 14g, where we should be aiming for 20g or more. Fibre comes entirely from fruit and vegetables and is divided into two kinds. The more familiar, insoluble form is the kind of fibre you find in wholemeal bread, nuts, seeds and cereals (and a bowl of *All Bran*). It acts as a kind of dietary drain brush – your stomach can't digest it, but it helps pull everything else along, keeping the system moving. This may be linked to the reasonably good evidence that cereal fibre in particular helps reduce the incidence of colorectal cancer.

The soluble fibre, which is found in other vegetables, some fruit and oats, barley and rye, can be digested and has the beneficial effect of reducing cholesterol levels (though to a relatively small degree). It can also help make the consistency of the material passing through the gut more malleable, in part by slowing the process. There is some evidence that all dietary fibre also helps reduce the risk of dying following a heart attack.

Generally speaking, some ways of getting fibre are pretty obvious – going for the more fibre-rich wholegrain and brown versions of starchy foods like bread, rice and pasta (or skin-on potatoes) is a great start. Less obvious is the fibre in other fruit and vegetables, but it's there. It is largely destroyed by putting the food through a blender – so while smoothies might be an easy way to contribute to your five a day, they are by far the worst way to consume fruit and vegetables.

LINKS:
- **Five a day** – page 47
- **Smoothies** – page 93

Five a day

**The broad concept of 'five a day' is good,
but the balance between fruit and veg
should be tipped towards the vegetables
– and the number 'five' is arbitrary.**

For some years now, UK government advertising and schools have been pumping out the concept of 'five a day' – the idea that you should eat at least five portions of fruit and vegetables a day. But there is a danger that the attractive simplicity of this mantra means that we don't do the best we can.

It's worth establishing what a portion is in this context. The recommended measure is that it's about the size of your fist, which cunningly adjusts the portion size between children and adults – though you have to be a little imaginative to see how, say, a banana fits. As an example, a portion is one apple, banana or pear, three heaped tablespoons of cooked vegetables or three pieces of celery.

Secondly, although the standard measure doesn't distinguish, ideally at least three of the five should be vegetables. Fruit is good for you – and you should eat some every day – but it is heavily loaded with sugar. Be particularly wary of smoothies, which both destroy much of the structural benefit of the fruit (fibre, for instance) and contain really eye-watering amounts of sugar.

Some vegetables have limited impact. Potatoes don't count at all, for instance, and pulses like peas and beans only count as one portion, however much you consume – though they are excellent foods and should make a frequent appearance. Fruit juice also counts as only one portion and, again, has a high sugar content. It doesn't matter if drinks say 'no added sugar' – these are natural sugars from the fruit, which are just as bad for you (see **Sugar**). Note that the 'five' part of the rule is arbitrary and not based on any science. Australia, for instance has a two (fruit) plus five (vegetables) slogan – in effect, seven a day. A US guideline from 2005 recommends between four and thirteen servings, depending on the individual's dietary needs.

Quite separate to the general advice, a large-scale study published in 2014 showed impressive decreases in risk of death thanks to eating more fruit and particular vegetables. Risk of death over the seven-year course of the study was reduced by 14 per cent with 1–3 a day, 29 per cent with 3–5, 36 per cent with 5–7 and 42 per cent for over 7. The study showed a clear progression from vegetables, providing the most benefit, then salad, then fruit, emphasising that more than half of your quota should be veg. Interestingly, fruit juice showed no benefit at all, and canned fruit juice had a small increase in the risk of death. The only proviso here, as with a lot of studies, is that it is possible that the kind of people who ate more vegetables and fruit might also be the people most likely to exercise and lead healthier lifestyles in general, as the study was not controlled for other factors.

Be wary of processed products claiming to be 'one of your five a day'. This has to be technically true, but it could be the least effective combination, and it doesn't make the product healthy, as it could easily be combined with sugars and fats. It is perfectly possible to make a packet of biscuits with fruit in them reach the 'one of your five a day' requirement, but that doesn't make them healthy. Be particularly suspicious of 'half of your five a day' as this has been used on a label to mean half of one of your five a day, rather than half your total five a day requirements.

LINKS:
- **Fibre** – page 45
- **Sugar** – page 102

Fizzy drinks

**Most of us enjoy a fizzy drink, but they are
often regarded with a lot of suspicion. What
are the implications of fizz for our health?**

Fizzy drinks are sometimes called 'carbonated', reflecting the fact
that the fizz comes from the gas carbon dioxide, partly dissolved in
the liquid. This technique dates back to the 1770s. Joseph Priestley,
the discoverer of oxygen, lived at one point near the Jacques brew-
ery in Leeds and experimented with the gas that bubbled off the
fermenting beer. He found if he bubbled it through water, some of
it dissolved, making the water taste like the sparkling mineral water
found in the Alps.

Priestley lacked the commercial drive to make a profit from
his invention, which was taken up (without payment) by the Swiss
Johann Schweppe – the rest, as they say, is history. What started as
a minor improvement on water became big business.

We know that sugary fizzy drinks are bad for us simply in the
quantity of sugar they contain – a can of cola, for instance, contains
a man's recommended daily amount of sugar, according to the latest
findings, or half that of the older recommendation. But what about
the calorie-free versions?

There was research published in 2013 that suggested that diet
drinks, despite being sugar-free, make you fatter. This was based on
a large US study. What this found is that overweight people who
switched to diet drinks tended to up their food calorie intake to
compensate. There is no suggestion that actually consuming a diet
drink will make you fatter – diet fizzy drinks compared to full-sugar
versions will reduce your calorie intake if you stick to a sensible balanced diet, so they remain a sensible option. (It's also worth saying
that US consumption of food and drink is quite different from that

in the UK and it is difficult to draw any conclusions from this study except about the US.)

A second concern from fizzy drinks is their impact on teeth, particularly for children. Clearly, sugary fizzy drinks have a bad effect on tooth enamel (as do fruit juices), but is there any impact from the fizz itself, which would still be a problem with diet drinks? There are plenty of stories to this effect online. The answer is yes … and no.

Sugar-free drinks, like diet cola, do not cause tooth decay in the same way sugary drinks do, by feeding the bacteria that cause damage. However, because carbon dioxide dissolved in water is a weak acid (carbonic acid), that acid does have a mild eroding effect on the teeth. This effect is worst with fruit-based fizzy drinks and colas, both of which have other acids as well, but it applies to all fizzy drinks.

The ideal would be, after enjoying a fizzy drink, to give your mouth a good rinse out with water to ensure the acids don't continue to attack your tooth enamel. Perhaps the most important thing is not to drink fizzy drinks – even diet drinks – or fruit juice after cleaning your teeth at night, when the drink will then have hours to work away at your teeth. It is also recommended that you don't put any kind of soft drinks or fruit juices in babies' bottles.

LINKS:
- **Artificial sweeteners** – page 9
- **Sugar** – page 102

Food colouring

There has been a lot of concern about food colouring, and in the early days there was considerable justification for this – but now things are significantly better.

Food colouring has been used since ancient times to make food look more appealing. Originally natural colorants were employed, like the vibrant deep yellow produced by saffron, the stigma of a type of crocus flower, or the purple-red cochineal, rather disgustingly produced from a parasitic bug found in South America.

Use of colouring is sometimes arbitrary – as when your curry turns up bright red – but it also has value in terms of taste and acceptability of foods. When margarine was first developed, for instance, it was white, but it was made more acceptable by dyeing it yellow, as this was the colour that spreads were expected to be. Equally, if you give someone a steak that is white or blue, it doesn't matter how much it tastes exactly the same as the normal version, it will seem different to the person eating it. Many people can't tell the difference between red and white wine if they are served blind at the same temperature – but the colour still influences their choice. (In effect, red wine is white wine with an added natural colouring from grape skins.)

When artificial dyes started to be produced in the 19th century, they were added to foods with more enthusiasm than concern for safety. Food manufacturers merrily piled in lead oxide for its red colour and used both a salt of arsenic and cyanide (Prussian blue) to make tea leaves look more appealing. Over time, though, we have weeded out the colourings that were dangerous, and these days the food colourings you purchase in the supermarket should cause no more problems than any other foodstuffs.

A good example of the way that colourings have been subject to checks and balances is the yellow colouring tartrazine. This bright yellow dye used to be found in everything from orange drinks and sweets to marmalades and cereals. A small number of people are allergic to the substance, but the widespread concern came from studies that suggested tartrazine, and a number of other additives, contributed to hyperactivity in children.

There are some problems with the studies, which weren't good enough quality to be sure of the conclusions. One problem is that hyperactivity is a very subjective condition, and it is difficult to systematically identify it and isolate its cause. Even in these studies, tartrazine was not shown to cause problems in isolation, but it was as part of a cocktail of additives, usually including a number of other colourings, that it did seem to have some effect. Although the evidence is weak, it seems sensible to minimise the use of tartrazine, and it is now rarely found in foods.

LINKS:
- **Chemicals** – page 21
- **E-numbers** – page 40

Food groups

All the evidence is that a good diet is one that makes appropriate use of the different food groups; so having a basic knowledge of them is helpful in understanding what's best to eat.

While some jokingly suggest the most important food group is chocolate, in practice the food groups bring together foods with similar nutritional properties to help make a sensible decision on what to have in your diet.

Although the division is somewhat arbitrary, the groups are usually fruit and vegetables, starchy foods, protein, dairy and sugary foods. The starchy foods, which are those high in carbohydrates, tend to be the 'staples' like bread, rice, pasta and potatoes. In the protein category come meat, fish, eggs, nuts, seeds and pulses like beans and peas. It's also worth saying that the final category isn't really sugary foods per se – fruit are very sugary foods – but rather foods that are unnecessarily high in sugar or fat – the bad boys. Like, yes, chocolate.

Broad advice for most of us is that we could do with more fruit and vegetables and less high-fat and particularly high-sugar food. (Because of the need to restrict sugar, that 'fruit and veg' should be more vegetables than fruit.)

Various methods are used to show the proportion of different food groups recommended. There is no absolute science to this – it's just a broad guideline – but they are often portrayed as a pyramid or a divided-up plate. The guidance from the US is currently to have about 30 per cent carbohydrates, 30 per cent vegetables, 20 per cent fruit and 20 per cent protein – plus an extra 15 per cent (don't ask me how they make this add up!) for dairy. Sugary fun does not get a look in.

The UK equivalent goes more for portions than percentages. It suggests plenty of fruit and vegetables each day (at least five portions), with a similar amount of starchy foods – preferably wholegrain. 'Some' milk and dairy (they seem to recommend two to three portions from their graphic), 'some' non-dairy protein and a small amount of high fat and high sugar.

LINKS:
- **Carbohydrates** – page 19
- **Chocolate** – page 25
- **Fat** – page 44

Fried food

**Recent evidence that saturated fats are not as bad
for us as we thought might make it seem as if it were
a green light for fried foods – but it isn't that simple.**

It now seems there is little difference in health risk between saturated and unsaturated fats, but this doesn't give free rein to eat all the fried food you like, whether you use butter, animal fat or vegetable oil.

Frying involves applying a fierce heat to the food, and the result is a number of physical and chemical changes. Water boils off and fat is absorbed by the materials being fried, giving it that familiar greasy texture. Most importantly, the fats themselves undergo changes, with some of the fats becoming trans fats, the worst of the dietary fats. This particularly happens if frying oil or fat is reused, which is why fast food can have unusually high levels of trans fats.

While eating fat in and of itself does not seem a significant issue for obesity or heart disease, it is still a source of calories, and there is no doubt that we can have too much fat to attain a properly balanced diet, which eating lots of fried food makes more likely to occur. But the trans fat content is the most worrying aspect of fried food.

Just as with the processed meat in bacon and sausages, it isn't the case that occasionally having some fried food is going to be a problem for your health, but it's best to keep it down to once or twice a week.

If you want to fry, a stir fry is the ideal approach, as only a small amount of oil is used and the food is in the oil only for a relatively short time, reducing the opportunity for the fat to soak in and minimising the amount of fat that is changed into trans fat.

LINKS:
- **Fat** – page 44
- **Processed meat** – page 82
- **Trans fats** – page 109

Fruit

It should be no surprise that fruit is good for you – but because it doesn't usually come with a content list, it is worth bearing in mind there's a whole lot of sugar in there.

Few of us eat as much fruit and vegetables as we should, so the last thing anyone should do is discourage you from eating fruit. However, it is worth bearing in mind that fruit contains a lot of sugar, and all the evidence is that excess sugar is probably the worst thing in our diet (see **Sugar**).

It is instructive to look at the sugar content of a couple of fruits because, unfortunately, food labelling does not extend to unprocessed foods. It would really help if it did, particularly as all fruit is not the same. Blueberries, which come up a lot here because they are often portrayed as a 'superfruit', are around 10 per cent sugar, bananas and mangoes 14 per cent, and grapes, up at the top of the list, are 16 per cent sugar. This would earn them an amber traffic light for sugar if they had UK-style food labels. If you want a low-sugar fruit, then go for the ones with a natural tartness – lemons come in at just 2.5 per cent, cranberries at 4 per cent and raspberries at 4.5 per cent. All these would get a green traffic light.

The good news if you eat whole fruit, rather than juice or smoothies, is that the sugar is less likely to damage your teeth, because it is locked away in the structure of the fruit – but it will still make up part of the sugar in your diet. The official UK recommended maximum is 70g/50g per day for men and women respectively, but a better target, given the dangers of sugars, is now considered to be 35/25g. A typical banana will have over half that amount of sugar – so if you are going for 'five a day' it is best to only have one or two of these as fruit.

LINKS:
- **Five a day** – page 47
- **Sugar** – page 102

Fruit juice

Like smoothies, fruit juice contributes to your five a day, but it is better to get your fruit whole if possible.

Many of us like a glass of fruit juice with our breakfast, and the good news is that it counts as one of your five a day (though only one, however much you have). But there are a couple of reasons why it's better, if possible, to have whole fruit than the juice.

As you can see in the **Fruit** section, fruit is an important part of our diet – and we definitely should try to have some every day – but because it is relatively high in sugar, we should limit ourselves to one or two portions of fruit, including juices and smoothies, per day. A basic juice lacks the fibre content of fruit – it's better to go for the juices with 'bits' in, which at least maintain some of the fibre. The other disadvantage of juices is that the structure of the fruit helps keep the sugar parcelled away until it gets to your stomach. A fruit juice gets to work straight away, providing sugar to the bacteria on your teeth, encouraging tooth decay.

You might think that this means it's a good idea to brush your teeth after drinking fruit juice, but it isn't. The acid in the juice weakens the tooth enamel, which can then be damaged during brushing. The recommended approach is to brush your teeth *before* a meal including fruit juice. That way you don't risk damage. However, a good rinse with water after a fruit juice is a good way to reduce the level of sugar that is left behind.

LINKS:
- **Fibre** – page 45
- **Five a day** – page 47
- **Fruit** – page 55
- **Smoothies** – page 93
- **Sugar** – page 102

G

· · · · · · · ·

GI

**Some of the information encouraging us to cut back
on carbs or balance our diets suggests we keep an
eye on the GI – but what is it, and does it help?**

You will quite often see 'GI' levels on packaging of food these days, which is supposed to help judge which foods are better for your diet. GI stands for 'glycaemic index' and is a measure of how quickly eating that particular food will increase your blood sugar level.

It is true that some of the best food as part of a balanced diet – fruit, vegetables, pulses and wholegrain cereals – all tend to be low GI, but this isn't a magic indicator that the food will be good for you (or help you lose weight). The good news is that if blood sugar levels change relatively slowly you will probably feel fuller for longer, but just because a food has a low GI does not mean that it is healthy. Chocolate cake, for instance, has a relatively low GI, while parsnips and watermelon, both far healthier, have a high GI.

Generally, GI is a better guide when applied to unprocessed foods than processed. This is because adding fat or protein to a starchy food will actually reduce its GI. This means, for instance, that crisps are lower GI than jacket potatoes, despite being significantly less healthy.

LINKS:
• **Carbohydrates** – page 19

Gluten

Gluten is a complex natural molecule that gives dough the stretchy, gooey texture that makes great bread. For most of us gluten is harmless, but a few are intolerant, and others believe a gluten-free diet would benefit us all. What's the truth?

Fewer than one in 125 of the population is gluten intolerant. They get stomach pains, constipation and diarrhoea, most often because they have coeliac disease, a genetic disorder of the small intestine where the immune system gets it wrong and thinks that the gluten is attacking it, causing inflammation. Rather more people claim gluten intolerance than actually have it – for the rest it is a 'nocebo', the negative equivalent of a placebo effect, where believing something is bad for you causes pains with no physical cause.

There are simple tests that your GP can perform to check if you do have a genuine gluten intolerance. If this proves to be the case, then it's a matter of ensuring that the diet is gluten-free, something that is increasingly easy with a wide range of gluten-free products available in the supermarket. In general, food should be fine as long as you avoid wheat, barley and rye. (Oats, while not a problem in themselves, are probably best avoided also as they are often contaminated with gluten from other grains.) Other cereals like maize and rice are fine.

There has been something of a fad diet of avoiding gluten – the kind of diet often endorsed by celebrities rather than experts – in those who don't have a gluten intolerance, probably on the assumption that people must be intolerant because there is something toxic in the gluten and that avoiding gluten must be 'good for digestive health'. In fact, it's the reverse – people who are gluten intolerant have a mutation that makes them react badly to a harmless substance that is a normal part of our diet.

The general argument for a gluten-free diet if gluten doesn't cause you problems is open to interpretation. Here's nutritional therapist Deborah Thackeray, quoted in the Morrison's supermarket magazine:

> Gluten doesn't need to be in sausages or many other foods in which it is often present. There is growing evidence that gluten is difficult to digest and while we need protein, the right sort of fats and plenty of vitamins and minerals, we don't need gluten. All of these nutrients can be found in other, non-gluten-containing foods.

This doesn't make a lot of sense as dietary advice – after all, fibre is hard to digest, but it's still good for you (see **Fibre**). All the evidence is that supermarkets are using 'gluten-free' rather as they do the 'organic' label: as a way to put a considerable mark-up on a product that doesn't cost them much more to produce. Gluten-free products can be a rip-off. Trials have failed to demonstrate any health benefit to a gluten-free diet for those who aren't intolerant, and it makes it significantly harder to get the fibre (and some vitamins) we all need. What's more, many gluten-free products are higher in fat than their normal equivalents, as something has to be used as a substitute to give the textural contribution of gluten. So, a gluten-free diet is best avoided if you don't have a medical reason to be on one.

LINKS:
- **Fibre** – page 45
- **Placebo effect** – page 315

GM foods

Genetic modification of foods could bring huge benefits, especially in countries where nutrition is poor, but misleading campaigning and corporate greed have made them a diet bogeyman.

Our reaction to GM foods, led by hysterical newspapers labelling them 'frankenfoods' and ill-founded campaigns by green groups, has caused this potentially life-saving technology to be thought of primarily as a risk. This is an example of being over-influenced by the feeling that 'natural' has to be better. Pretty well *all* the food we eat is genetically modified. If you doubt that we've been busy with genetic modification for thousands of years, consider a few simple examples.

First take a look at a Chihuahua and a Great Dane alongside each other (admittedly not food in most places, but a very visible example of GM at work). Both dogs were bred from the same basic wolf stock – but the modification is extreme. Then take a look at sweetcorn (maize) and a cauliflower. Sweetcorn has been so modified from wild maize that it is no longer possible for its huge seed heads to propagate. It can only reproduce with human intervention. Cauliflower is a mutant cabbage, where the flowers have grown out of control and fused into a useless, if tasty, lump – again, it can't breed naturally.

Despite reports you will see to the contrary, there is no evidence of GM food causing risks to humans. For instance, a study that is commonly quoted was supposed to show that rats fed on GM corn developed tumours and died prematurely. The study in question was totally without value, making every technical mistake in the book. It was retracted, meaning it was unsupportable, yet it continues to be quoted as scientific 'fact'.

This doesn't mean we don't have to be careful with GM. There are dangers with introducing elements into a genome that will spread to the wild. Yet the dangers of GM have been hugely overstated – providing the technology is used responsibly. It doesn't help that early on, seed companies tried to use GM to lock farmers into buying their seed and the matching herbicide – but that doesn't mean there is anything wrong with GM itself, just the way it was marketed.

The whole GM debate is complex enough to fill a book in its own right, but it has certainly been subject to the same misuse of information from both green organisations, which often oppose it on principle without thinking about it in detail, and from tabloid newspapers. A genetically modified variant of rice that was designed to counter vitamin A deficiency was dismissed by Greenpeace because the environmental organisation said that to obtain the required amount of vitamin A would require 'seven kilograms a day of cooked Golden Rice'. The actual amount is 200 grams.

Major developments are under way at the moment to produce plants that are more able to resist the impact of global warming – drought resistant strains of maize, for example, or rice that is better able to cope with higher salt levels, as evaporation leaves water more saline. As always there are genuine concerns – for example, it's possible that the salt resistance could spread to wild rice variants that could then clog up estuaries and disrupt the natural ecosystem. But it may be, as climate change proceeds, that we will have to take more risk with GM, just as we have in the past with 'natural' genetic modification. And certainly, if we continue to resist GM to the extent we currently do, we have to do so aware that the consequences will include many more deaths in the developing world.

H

. . . .

Hydration

**Water is essential for life, but it can come
in many forms, and the suggestion that we
should drink 'eight glasses a day' is a myth.**

Our bodies are, on average, around 60 per cent water, and everything from our digestion to the operation of individual cells would
not function without it. It's not surprising that we can only go
around three days without drinking water, as we are constantly losing it and need to keep levels topped up to avoid dehydration. But
the good news is that keeping hydrated is nowhere near as difficult
as you might have heard. (Just as well, when you consider how many
millennia we've managed without sports drinks.)

It is often said that you need to drink eight glasses of water a
day – around 2 litres or 3½ pints – but there is no scientific basis for
this. A fair amount of research has been done to try to track down
where this persistent myth came from. It's possible it was from a
1945 US National Research Council recommendation that adults
should have a millilitre of water for each calorie of food.

Two litres is, indeed, a typical required amount of water if you
aren't exercising so hard that you sweat in large quantities, but we
don't need to *drink* this amount. We get about half of it from our
food. As for the rest, Heinz Valtin, a kidney specialist from the
Dartmouth Medical School in New Hampshire has reviewed the
available evidence and shown that there is no need for this to be
pure water. Tea and coffee, for instance, are fine (provided you aren't
drinking so much that there are problems with caffeine). You may
have heard you should avoid these (and colas) because they are
diuretics, which make the body lose more fluid than it gains – but a

2000 study showed there was no difference in the effect on hydration levels between water and caffeinated drinks.

Even high-water-content alcoholic drinks like beer, if not taken to excess, contribute to hydration. As far as other innocuous drinks like milk, squash and fruit juice are concerned, as long as you keep your eye on sugar and calories, again, they are just as good as water for hydration. Sports drinks work fine too, but research in the *British Medical Journal* showed that, despite claims from drinks manufacturers to the contrary, thirst is a fine guide as to when you need to hydrate. There is no need to 'stay ahead' of your thirst – you should wait until you are thirsty. And there is no evidence we need to replace electrolytes, the substances that support the body's electrochemical mechanisms, in our drinks – these are replaced naturally from what we eat. Sports drinks are no more effective than water at hydration.

Sports enthusiasts should also be particularly wary of drinking water too quickly. Gulp down several litres at a go and you will reduce the electrolyte levels in your body to the extent that cells swell and the brain can be affected, in extreme cases leading to death. It is sobering to note that several marathon runners, who are particularly at risk as they are deluged with opportunities to drink over the course of the run, have died from drinking too much. None has died from dehydration. Hydration also has no preventative effect on heatstroke, which is why some runners seem to mistakenly over-hydrate.

LINKS:

- **Alcohol** – page 5
- **Coffee** – page 29
- **Sugar** – page 102
- Hydrogenated fat – see **Trans fats** – page 109
- Margarine – see **Spreads** – page 97

M

• • • • • • • •

Monosodium glutamate

Monosodium glutamate, commonly used in Chinese cooking, is often blamed for a range of health problems, but there seems little reason to be concerned about it.

Monosodium glutamate (MSG) is a flavour enhancer. It doesn't taste of anything itself, but it makes umami, the fifth standard flavour sometimes called 'savoury', more pronounced. MSG is most famously associated with Chinese food, but it is used in most Asian cuisine, and particularly in Japan, where it was extracted from kelp long before it was known exactly what it was.

Back in 1969 a paper claimed that a number of people in the US were suffering from 'Chinese Restaurant Syndrome' where they experienced chest pains, headache, sweating and facial swelling or numbness after eating a Chinese meal. It was assumed that MSG was the cause, but the study has never been repeated, and there was no actual evidence suggesting that MSG was behind the response. Since then a number of people have claimed, again with no evidence, that MSG can be responsible for everything from obesity to brain damage and death.

It's puzzling that MSG was blamed in the first place, as glutamates are very common in foodstuffs and there is no suggestion that these have any negative effects. In fact, your body produces glutamates, which are involved in the energy-production cycle, in controlling waste nitrogen and as a neurotransmitter employed in communication by your nerves. The chances are that without glutamates in your body you would die.

One of the lessons we have learned from antioxidants and other supplements is that just because something is good in our body, it doesn't mean it's good to consume it. But having said that, the wide occurrence of glutamates in other things we consume makes it extremely unlikely that MSG has any negative effect. For instance, there are higher concentrations of glutamates in soy sauce (and for that matter in tomatoes) than is ever used as a flavour enhancer. And there are huge concentrations in Marmite/Vegemite and Parmesan cheese. It's entirely possible that Italian food, rich in tomatoes and often using Parmesan, has a higher glutamate content than Chinese food.

To confirm the apparent safety that emerges from the logic of consumption, there has now been a series of reviews and properly controlled clinical trials that have shown that MSG has no negative effect. It is entirely safe to consume.

LINKS:
• **Antioxidants** – page 7

N

· · · · · · · ·

Natural products

**Advertisers and marketing people use
words like 'natural' all the time as they
trigger a knee-jerk reaction. But we need
to be more than a little careful.**

It is difficult to avoid the word 'natural' in advertising. This is because it is a word that immediately triggers a whole collection of nice, warm feelings, making any product that is labelled as 'natural' seem more attractive.

This is understandable. Think, for instance, of the difference between the natural and the built environment, between a natural animal and a robot. Almost all of the alternatives to 'natural' sound unpleasant. 'Unnatural', 'artificial' and 'chemical' are all used in a derogatory fashion. But they are just words, and we need to hold back from the truly emotional response, because the natural option often isn't the healthy or safe one.

Part of the problem here is that we tend to forget that many poisons are natural. Think, for instance, of the many deadly toadstools or the incredibly poisonous castor bean plant, which contains the most toxic substance known, the neurotoxin ricin.

'Natural' is often used to get money out of us. It might be that only a small and irrelevant part of a product is actually natural. Look at the contents list of practically any shampoo, for instance, that uses words like 'natural' or 'herbal' or 'country' to give this lovely warm glow, and the main constituents are likely to be exactly the same soap or detergent chemicals as any other shampoo.

Another way that 'natural' is used to make money is by selling us natural-looking rubbish. Many people would rather buy carrots with their green tops on and covered in soil than they would carrots that have been trimmed and washed. Surely those earthy carrots are closer to nature – more like the moment they were pulled from the ground? But this makes no difference to the taste. Instead, you are paying for the weight of the stalks and that plastering of soil – which you will then throw away when you do the job that has already been done for you with a clean vegetable.

A natural product can be very pleasant – but remember that nature at its worst is extremely unfriendly. Bacteria and viruses are natural and may well be present on food that hasn't been properly washed. 'Natural' when applied to a product is always a marketing label. Don't be taken in. Appreciate what's good about nature – but remember as you sit in your unnatural house on your manufactured furniture, keeping nice and warm thanks to your artificial heating system, listening to the natural wind and rain crashing against your unnatural windows, that nature is sometimes best kept at bay.

LINKS:
- **Psychology** – page 211

No added sugar

Just as marketing people use weasel words like 'natural' to persuade us a product is good for us, we need to be wary of the 'no added sugar' label.

It is easy to feel reassured when we see the words 'no added sugar' on packaging. We can feel safe buying this product, particularly for our children, because it is clearly good for them. But that isn't clear at all.

All 'no added sugar' tells us is that the product has less sugar than the same product 'with added sugar'. So, for instance, anything fruit-based will have sugar in there already. There is no need to have added sugar to make it a sugary product. Various other natural products have a surprising amount of sugar (12.5g in a 250ml glass of skimmed milk, for instance).

So, in and of itself, 'no added sugar' is not the health green light it might seem to be. Of course, if you have a choice between a product with no added sugar and the same product without that label (which probably means it does have added sugar) – squashes are a good example – then it's well worth going for the 'no added sugar' version, but don't assume it means it's sugar-free. Always check the actual amount of sugar present.

LINKS:
- **Artificial sweeteners** – page 9
- **Sugar** – page 102
- Use of 'natural' in marketing – see **Natural products** – page 67

Nuts

Although nuts are a relatively fatty food, they are far better to snack on than crisps or chocolate and have positive health benefits too.

Nuts have a long record of being identified as healthy. This might seem counter-intuitive because nuts are high in calories, being up to 50 per cent oil. In 100g of nuts you will typically find a huge 600 calories (that's more than a Big Mac). Yet all the evidence is that regularly eating nuts is a positive benefit for weight loss. A 2007 study showed that frequent nut eaters gained less weight and were significantly less likely to be obese than non-nut eaters.

While there is always a risk with such links that another factor is causing both observations (for instance, people with generally healthy lifestyles are less likely to be obese and are more likely to eat nuts), there is some sense to the observation. If you are snacking on nuts, you are less likely to eat chocolate and crisps, and the makeup of nuts seems more likely to suppress hunger than many snacks.

As far as positive health benefits go, a very large study reporting in 2013 made it clear that regular nut eaters were likely to live longer, particularly having a lower death rate from heart disease, cancer and respiratory diseases. The study found that those who ate nuts (specifically almonds, brazils, cashews, hazelnuts, macadamias, peanuts*, pecans, pine nuts, pistachios and walnuts) at least once a day were 20 per cent less likely to die in the survey period from 1980 to 2010, and those who ate nuts two to four times a week still had an 11 per cent reduction in chance of death in that period.

What it seems to come down to is that nuts have more of the usual suspects for health benefits from food and fewer of the bad varieties. We might once have pointed out that they are good sources of unsaturated fat and omega-3 fatty acids, with less saturated fat than olive oil. Of itself, given the latest research, this doesn't give them an advantage, but throw in good dietary fibre, cholesterol-lowering plant sterols and excellent micronutrients like magnesium and copper, folic acid and vitamin E and you have an impressive bunch of beneficial ingredients.

Pistachios, and to some extent peanuts, also give us resveratrol, for which there is some uncertain evidence of anti-ageing benefits.

If you don't like nuts, you probably won't manage enough of them to derive much benefit. And of course this is not an option

---

* Pedants will point out that peanuts aren't actually nuts, but are actually legumes that are closer to peas, but groundnuts like peanuts were found to have the same benefits in the research.

open to those with nut allergies. But if you do like eating nuts – and most of us do – they are easy to add to salads and are tasty and convenient to snack on. Getting in around 30–50 grams a day is really no hardship. So, why not stock up on nuts? (Avoid ready-salted to keep salt consumption down.)

LINKS:
- **Fat** – page 44
- **Salt** – page 90
- **Omega 3** – page 72

O
. . . .

Omega 3

**We are used to being bombarded with claims that
food products and supplements contain omega 3 –
but what is it, and what benefits does it actually give?**

Omega 3 is a group of fatty acids, naturally occurring chemicals
with names like all-cis-9,12,15-octadecatrienoic acid. The name
refers to the bonds, the way the elements in the chemical are joined
together. In an omega 3 fatty acid, the first double bond, where two
carbon atoms are doubly connected, turns up three positions away
from the end of the molecule ('omega', the last letter in the Greek
alphabet, is used to designate the end of the molecule).

Omega 3 fatty acids occur in some kinds of oil, notably fish
oils and also the oils of some nuts and seeds (walnuts and flax
seeds do particularly well). A wide range of health benefits have
been ascribed to them over the years, particularly stabilising blood
pressure and levels of triglycerides, a fat in the body that is linked
to heart disease when in excess. They are also considered beneficial to women who are pregnant or breastfeeding, as they help a
baby's nervous system develop. Some claim they help with arthritis,
combatting depression and more. The most extreme claims are for
improvements in intelligence and for resisting the tendency to have
reduced mental abilities as we get older.

Of these claims, the brain-associated ones have the least scientific
support – in fact, there is no significant evidence of any benefits there.
On heart-related issues, a comprehensive review of studies concluded:

> It is not clear that dietary or supplemental omega 3 fats alter
> total mortality, combined cardiovascular events or cancers

in people with, or at high risk of, cardiovascular disease or in the general population. There is no evidence we should advise people to stop taking rich sources of omega 3 fats, but further high quality trials are needed to confirm suggestions of a protective effect of omega 3 fats on cardiovascular health.

A 2014 survey of 72 studies found that there was no significant evidence that omega 3 (or the related omega 6) polyunsaturated fats protect the heart.

The best advice is that while there are no proven health benefits, there is no evidence that omega 3 is bad for us, and foods containing omega 3 are generally good for us (bearing in mind the warnings below) – but these fatty acids are not the universal good guys they were once believed to be.

There are a couple of warnings if you decide to beef up on omega 3. Fish liver oil supplements tend to be high in vitamin A, one of the few vitamins it is possible to overdose on – so check your dosage. The best sources if you eat actual fish are the oily fish like mackerel, salmon and sardines, or some shellfish, like mussels and crab. Tuna is an oily fish, but only fresh tuna maintains its fatty acid levels. There are recommended maximums for oily fish consumption of two portions a week for girls and women who might become pregnant one day, or who are currently pregnant or breastfeeding, and four portions a week for the rest of us. This is because oily fish store up some pollutants like dioxins and PCBs, for which we want to limit intake.

LINKS:
- **Breast milk** – page 266
- **Fat** – page 44
- Vitamin A – see **Vitamins and minerals** – page 111

Organic food

**Organic food is often held up as the pinnacle of
food quality. It does provide a number of benefits
– but not always the ones you might expect.**

As soon as we hear 'organic', we think of healthy food, produced
the traditional way without chemicals and pesticides. It feels like
it should be tastier and healthier, with better animal welfare, and
should be better for the environment. We'll see in a moment how it
stacks up against these measures, but the first thing to bear in mind
is that 'organic' is being used as a marketing term. 'Organic' has a
very specific meaning in science, referring to something based on
chemicals containing complex carbon compounds. So, for instance,
you can't have 'organic salt' (though I once bought some) – salt is
inorganic. But in marketing speak, organic means grown or raised
according to certain standards.

In the UK, these standards are primarily set by the Soil
Association. One thing to be wary of is that other countries have
wildly different organic standards, and may label food organic that
the UK would not consider to be so, for instance in terms of animal
welfare. Only with British or American reared and grown organic
produce can you be sure that it is truly organic. But supermarkets
are happy to sell anything they can give an organic label to, as from
their viewpoint it is simply a way to make extra profit, because they
can charge more for organic goods.

There are problems with organic standards. The movement
began in a very non-scientific way, more mysticism than any real
understanding of what was going on in an agricultural process.
Because of this, the organic regulations can sometimes make the
environmental impact or animal welfare worse than conventional
farming. For instance, until recently the Soil Association liked the
use of copper sulfate as a fungicide because it was 'traditional',

even though it is much more environmentally damaging than other fungicides, and insists that organic animals should be treated homeopathically, sometimes resulting in considerable suffering.

Taking a look at the potential advantages of organic foods, there is no taste benefit whatsoever. Where organic food is locally raised, fresh etc. this can make for an improvement in flavour, but exactly the same goes for locally raised, fresh, non-organic food. The same goes for the nutritional benefits of organic food – there are none. Reports of a 2014 study appeared to show otherwise, but in fact the study did not show any significant benefits, as it referred to the unsupported benefits of antioxidants. Those reporting the survey often failed to mention that the only true nutritional benefit discovered was that non-organic food tended to have a little more protein than the organic equivalent. There are also no health benefits from the composition of the food. Helen Browning of the Soil Association once told me: 'The only health claims I make for my organic bacon is that it is healthier than eating a bag of doughnuts.'

One reason given for eating organic food is to avoid the pesticide residues left on produce by conventional agriculture. We know that some pesticides are dangerous, and many conventionally farmed products do retain some residues. Joanna Blythman from the Soil Association made a strong case for the dangers of residues in our food, saying: 'You can switch to organic, or you could just accept that every third mouthful of food you eat contains poison. Are you up for that?'

This is entirely incorrect. Practically *every* mouthful of food we eat contains a whole range of poisons, some natural, some synthetic. It certainly makes sense to wash fruit and vegetables before consuming them. This applies equally well to conventional and organic produce. Organic carrots, for instance, may not have any pesticide residues, but they are certainly covered in soil, containing a wide range of microorganisms, probably including those responsible for tetanus and MRSA.

But we also have to be aware that whether or not something is poisonous depends on the quantities consumed. Even with the worst possible pesticide residues, the food we eat typically contains far more (just as dangerous) natural pesticides that are part of plants than the artificial variety. We typically consume 10,000 times more of the deadly natural pesticides than we do of equally deadly synthetic pesticides. Luckily, even that volume is not high risk, but it puts the levels of residue we consume on non-organic food into stark proportion.

One area where we can see real benefits from organic agriculture is in animal welfare (as long as the animals haven't been treated homeopathically). Organic standards are very good on animal welfare. British animal welfare in general is high in worldwide terms, but organic standards push things even higher and rival the best non-organic free-range farms.

And surely organic methods are also good for the environment? Here the argument is more nuanced. Non-organic farms make use of fertilisers and pesticides that have a negative impact on the local wildlife and make a contribution to global warming. However, organic methods are a lot less efficient and result in greater tilling of the soil, which results in an overall impact on greenhouse gas emissions that is slightly worse than conventional farming.

Overall, organics do have some benefits, but they are matched by local, fresh, free-range non-organic produce. You pays your money and you takes your choice.

LINKS:
- **Antioxidants** – page 7
- **Homeopathy** – page 295

P

. . . .

Paleo diet

**Every now and then a diet comes along that
tries to get us back to nature – to eat the
foods our body is 'designed for'. But eating
like cave dwellers is not the answer.**

Sometimes the reasoning for using a particular diet seems to make
a lot of sense. Human beings came into existence around 200,000
years ago and haven't evolved on a large scale since. For most of
those 200 millennia our diet was quite different from our modern
intake. So wouldn't it be a good idea if we went back to the kind of
things they used to eat – to what has been called a 'paleo diet'? After
all, the outbreak of obesity, heart disease, diabetes and some cancers
could well be connected to our 'unnatural' diet.

The idea, then, is to return to the diet of the hunter-gatherer. We
don't have to avoid meat, but should stick to game animals: deer,
wild boar, game birds and the like. Fish is fine, as are fruit, some
vegetables and, of course, nuts and berries. But other meats, dairy
products, wheat, corn and other grains, peas and beans, sugar and
oils go out of the window. A couple of areas are borderline. The
paleo diets tend to exclude eggs and salt, but both of these would
tend to be in the real 'cave dweller' diet, though we'd be talking about
scavenged wild bird eggs, rather than chickens, and though rock salt
was occasionally found and treasured, salt would mostly come from
consuming animal blood (so game-based black pudding might be
the answer for the purist).

Unfortunately, there are a couple of problems with this picture
of returning to nature. Firstly, just because this was what people ate
back then does not mean that the diet from that period is what was

best for them or will be best for us. It's just what was available, but it doesn't make it ideal. There is no magic match between what is readily available and what is good for us.

Secondly, while it's true that Homo sapiens has not evolved at a gross level, it is also a misleading statement. We *aren't* genetically identical to the original humans. And some of the changes that are common – such as the ability of most adults to consume milk – have a direct impact on what makes for our best diet. (The same goes for a genetic change that makes us better at digesting grains.) It doesn't make sense to attempt to reproduce the diet we 'evolved for', when we've actually changed to deal with some aspects of a modern diet.

The hunter-gatherer diet, combined with the lifestyle, does seem to be one that is reasonably good at keeping a population healthy, at least to breeding age, but that doesn't mean it's the best. And, of course, we can only get a partial picture of what they *did* eat back then, though one thing we know for certain is that almost all sources of nutrition, even game and vegetables, have changed more (in part through selective breeding) in the last 200,000 years than we have; so in going back to such a diet, we are actually feeding the less evolved humans with more evolved food – which shoots holes in the logic for adopting the diet in the first place.

It's even worse for those who believe we should confine ourselves to raw vegetables that could be found wild. This isn't in any sense our 'natural' diet – there is good evidence that humans have always been omnivores – and to consume a good balance requires spending hours a day eating, plus some supplements to deal with the dietary omissions from sticking to such food. Not recommended.

LINKS:
*   **Diet** – page 1

Pesticide residues

**It's not uncommon to worry about pesticide
residues on food, but in reality the risk is very low.**

Pesticides are harmful to human beings – some with very unpleasant effects (in fact, some chemical warfare products were derived from pesticides). And it is certainly sensible to wash all fruit and veg before eating it – not just because of pesticides but also because of very natural risks from bacteria that can cause food poisoning – as was shown all too clearly in Germany in 2011 when a number of deaths were caused by E. coli in salad that had not been washed.

However, pesticide residues are normally at very low levels, and like all toxins, the degree to which they are harmful relies on the dosage. A classic example of misunderstanding risk from pesticide residues causing a negative effect on health came from the Alar scare in the US in the 1980s. A chemical called daminozide, with the brand name Alar, which was sprayed on apples, was withdrawn because it was found to be carcinogenic when fed to lab animals in large quantities.

Of itself this isn't particularly surprising. There are plenty of substances we consume regularly like coffee and alcohol that also cause cancer when fed to lab animals in large quantities and we happily get through much larger quantities of these than the trace residues of Alar that may have been on apples. However, the risk from Alar was picked up by the major US TV show *60 Minutes* and campaigned against by high-profile individuals including Meryl Streep (a key *Science for Life* red flag is whenever a celebrity gets involved). The result was that apple sales, and particularly apple consumption by children, collapsed.

Yet the research on which this panic was based depended on feeding animals 266,000 times the quantity of Alar a human might be exposed to, a problem compounded by the bizarre assumption that a toddler drinks 31 times the apple juice relative to its body-weight as does an adult. This number was just plucked out of the air. Result – the health of American children was worsened because they weren't eating anywhere near as much fruit.

The most important thing to realise when it comes to pesticide residues is that they make up only a tiny amount of the cancer risk of the food and drink we consume. The fact is, if you look at the risk from the quantities of poisons we consume, the last thing we need to worry about are pesticide residues. Looking, for instance, at the cancer risk from the average diet, 93 per cent of the risk comes from alcohol and 2.6 from coffee. Once we get the relatively danger-ous natural sources of risk like lettuce, pepper, carrots, cinnamon and orange juice out of the way, the first chemical contaminant is a chemical called ETU at 0.05 per cent. If you add up all the major chemical contaminants and pesticides at legal levels they have a similar risk to eating celery.

LINKS:
• Residues on food – see **Organic food** – page 74

Probiotics

**There is good evidence of benefits for probiotics
in helping with very specific medical conditions,
but not for the general consumer.**

There is a huge market for probiotics, worth around £20 bil-lion worldwide and £3–4 billion in the UK alone. A probiotic is a substance that encourages organisms to grow, and is generally used in these products specifically in the context of encouraging microorganisms – so-called 'friendly bacteria' – often including a significant load of the bacteria in the product.

An overview taken in 2008 of probiotics did find that they had value for certain very specific disorders. There was strong evidence that they helped manage acute and antibacterial associated diarrhoea, and there is slightly weaker 'substantial' evidence that they are a benefit with atopic eczema, a skin condition that is particularly common in infants. However, there was no evidence that they contributed to a normal healthy lifestyle.

The idea behind probiotics is the indisputable one that our bodies are occupied by many billions of bacteria that make a positive contribution to our digestion and other parts of the body's ecosystem. On a simple count of cells, your body is actually mostly bacteria, with ten times as many (around 100 trillion) bacterial cells as there are human cells.

It is possible to live without your friendly bacteria, but there is no doubt that they offer great benefits, both by helping the gut with digestion, particularly of plant matter, and also because they reduce the opportunity for bad bacteria to take hold by occupying pretty well all available surfaces.

However, the living conditions are inevitably difficult for bacteria in your gut, and it is only because of special protection systems that they aren't mostly eaten by the stomach acid or washed away. Most bacteria added in probiotics will simply be consumed or flushed through because they don't have this protection. If you want to encourage good bacteria, you are better off eating a high-fibre food like porridge oats, as they seem good at encouraging your resident friendly bacteria to thrive.

With the exception of the illnesses mentioned above, the only benefit from probiotics seems to be psychological. They trigger the placebo effect, which does have some mild beneficial effects for the body, and certainly helps you feel better. Of course, many probiotics are pleasant to consume. If you enjoy, for instance, a daily pot of yoghurt drink, then there is no reason at all not to continue enjoying it. But don't expect it to make a lot of difference to your health.

LINKS:
- **Placebo effect** – page 315

Processed meat

Fitting with the general feeling that 'everything we like is bad for us', those old favourites like bacon and sausages are proving controversial when it comes to dietary problems.

It's easy to get the impression that dietary experts are against fun, because they seem to have it out for everything that is tasty. But there's a good reason that sweet, fatty and salty things attract us – because they were once scarce and helped boost the health of someone near starvation. In our modern, everything-on-tap world they become a menace.

One example with a double whammy is processed meats like bacon and sausages – the staple of a traditional fry-up. 'Processed meat' also includes salamis, ham and similar cured meats. Not only are they usually high in fat and salt, but the processing they undergo has the potential to increase health risks. Dangers include increased risk of heart disease and some cancers, notably bowel cancer.

The main warning of health risk comes from a large study carried out across Europe looking at death rates over twelve years in a wide range of people – around 450,000 in total. It came up with a scary-sounding finding – that people who ate the most processed meat (averaging 160g or more a day) had a 44 per cent increased risk of dying in the following twelve-year period than people who ate the least (averaging 20g a day). This suggested to them that if we kept down to 20g a day or less (a small slice of bacon), then 3.3 per cent of deaths *caused by eating processed meat* could be avoided.

Unlike a lot of trials on diet, this was in part controlled for other factors. What this means is that we can to some degree take out the added risk that comes from, for instance, smoking. It could well be that the 160g a day people (that is, for instance, those who eat at least two sausages and two rashers of bacon every day) also were more likely to smoke, less likely to exercise and so on. I don't know if this is the case, though poor diet often goes hand in hand with other poor health habits. Though the people who ran the study tried to eliminate the impact of smoking, age, alcohol consumption and exercise levels, it is notoriously difficult to do.

However, this was still a 'cohort study' – it wasn't a proper scientific controlled trial where, for instance, there is a blind, randomised comparison of people who do and don't eat bacon. The evidence has to be called into question to some degree by the fact that most of the data was collected using self-administered questionnaires, and people are notorious for fibbing on these. (Have you ever been asked how many units of alcohol you drink a week? The majority of people who do drink alcohol give an inaccurately low figure.) And the researchers admit there still could be other health and lifestyle factors they didn't take into account that could be the real cause.

Overall, the study concluded there was a 'moderate association' between consuming more processed meat and dying earlier, and confirmed the usual advice that a balanced diet with plenty of fruit and vegetables is good for you. However, it would be reading far too much into this to suggest that it is necessary to cut out all processed meat (even the best performing group were typically eating the equivalent of around one fry-up a week) – the odd bacon sandwich or full English isn't going to do any harm. But it does make sense to keep consumption down and not indulge every day.

LINKS:
- **Fat** – page 44
- **Salt** – page 90

Protein

One of the essential food groups, protein is found in everything from meat to beans – but how much should we have in our diet?

Proteins are large, messy organic molecules that have a huge array of important functions in our body. They are made up of building blocks called amino acids (the genes in our DNA are instructions on how to combine amino acids to make different proteins), and we get these essential amino acids from consuming protein in our diet – so it is an essential part of what we eat. (We can also derive energy from protein if starving, though typically the energy-production process uses our body's own proteins.)

One of the principal food groups, the protein in our diet largely comes from meat, fish, eggs, dairy products and pulses like peas and beans. Some also comes from nuts and, to a limited degree, whole grains. Guidance on quantities of protein in a balanced diet tends to put it at around 20 per cent, or two to three portions a day.

Meat and dairy-based proteins have tended in the past to raise dietary concerns, both because they tend to be higher in fat than other protein sources and because, for instance, red meat has been associated with various health risks. However, with more evidence, we can now be happier that a sensible amount of fat and red meat presents no health risks and so we can enjoy the occasional steak or burger (from good quality meat) as one way to get protein without concerns.

It is perfectly possible for a vegan to get sufficient protein in their diet, primarily from pulses and their derivatives like the soya bean curd tofu, but it is generally recommended that babies and toddlers, who have a particular need for protein to help them grow and develop, have some of the other sources of protein in their diet.

LINKS:
- **Dairy** – page 32
- **Fat** – page 44
- **Food groups** – page 52
- **Red meat** – page 86

R

· · · · · · · · ·

Red meat

**Like processed meat, red meat is often held up
as a dietary villain – but the more data that is
available, the weaker the link seems to be. No one is
suggesting you eat steak every day, but it's nowhere
near as bad for us as is still often suggested.**

We like our dietary advice to be like the old Saturday morning
Western films. There you knew that the cowboys wearing the black
hats were baddies, and those wearing white hats were goodies.
Similarly with meat, for a fair while we've had the feeling of 'red
meat = bad; white meat = good'. (By red meat, we usually mean
beef, lamb and pork, but not the reddest of them all, venison.) The
reality, as is often the case, is rather more complex.

When the first studies were done, which typically lumped
together red meat and processed meat, there was thought to be some
evidence that red meat presented a similar, but slightly weaker risk
than processed meat. But the more evidence that was gathered, the
less strong this association became, to the extent that by the time the
450,000-person European study (referred to in the **Processed meat**
section) was done, they could detect no linkage in risk of cancer.
And there have even been studies that show a slight decrease in risk
of heart disease from sensible red meat consumption.

One of the big problems the people doing these studies face is
removing the impact of other factors, particularly the ones known to
be correlated with red meat consumption that we know are bad for
health. So, for instance, in the big European study only 13 per cent
of those in the lowest red-meat-eating male category were smokers,
where 40 per cent of the highest red-meat-eating males smoked. The

alcohol consumption was 8.2g per day for the lowest meat eaters but 23.4g per day for the highest meat-eating group.

To remove the impact of these known health hazards, the scientists had to make corrections. They changed the data to what they thought it should have been if these factors weren't involved. Unfortunately this inevitably involved a fair amount of guesswork. It is possible to make different, still valid, assumptions and produce outcomes that vary by 100 per cent or more.

The picture with red meat seems to be that our best evidence says you should be perfectly fine eating red meat on a regular basis, but it is best to avoid eating it every day, and in the kind of huge quantities found in US steak restaurants. There is no evidence that reasonable red meat consumption – up to around 700 grams a week – has *any* negative effect on your health. As always with food, a good balanced diet should involve a mix, rather than eating the same stuff day after day.

LINKS:
- **Fat** – page 44
- **Processed meat** – page 82

Red wine

**If there is one thing that has had a confusing story,
where we are repeatedly told that it is good and
bad for us, it's red wine. The truth is, it's both.**

Certain newspapers seem to be on a mission to go through every single food and drink and tell us whether it helps prevent or cause cancer. But red wine throws them into a spin. The *Daily Mail*, for instance has told us both 'Red wine may beat breast cancer' and 'Breast cancer could be triggered by just one glass of wine a day'.

The problem with the attempt to put something like red wine into a box labelled 'good for you' or 'bad for you' is that it isn't a medicine. It isn't a single substance with a controlled dose. It's a whole complex mix of things, some good for you, some bad, which vary considerably in impact depending on how much you consume and how regularly.

Alcohol, for instance, while not too bad in small quantities, is certainly not great for your health. It's by far the most carcinogenic substance in our regular diets, increasing the probability of getting a number of cancers. In excess it can also damage your liver, increase the risk of diabetes and heart disease and plenty more. But there are other things in there too.

For a while all the talk was of antioxidants, but we now know that these have little if any health benefit when consumed. We do know that there is something in moderately consumed red wine that does seem to reduce the risk of heart disease, reduces 'bad' cholesterol levels and prevents blood clots. The most likely beneficial substances are flavanols, which come from the grape pips, and tannins. (Both these substances are also found in teas.) Another substance called resveratrol originating in red grape skin is often considered to be the magic ingredient – suspected by some to have anti-ageing properties – though tests on the compound isolated from wine have had mixed results.

It is far too early to know for certain if there are significant health benefits for these various interesting compounds that crop up in red wine. They could indeed be good for us – but what we do know is that drinking red wine is not the best way to get them, because of the negative impact of alcohol. Most of these substances occur in other foods and drinks (sprouts, for instance, are also a good source of one of the flavanols, while resveratrol is more abundant in red grape skins than wine) with fewer negative side effects.

The message is, if you like red wine, enjoy it in moderation, bearing in mind that there is a risk from the alcohol. But don't think of it as a medicine.

LINKS:
- **Alcohol** – page 5
- **Antioxidants** – page 7
- **Tannins** – page 108

S

· · · · · · · ·

Salt

**Salt is a mineral that we need in small
amounts in our diet – but like many
substances that used to be rare, it is now too
abundant and can damage our health.**

Salt is one of the few inorganic substances you will find in the
kitchen cupboard, being the simple chemical compound sodium
chloride. Adults need around one gram of salt (half a level teaspoon)
a day, and should not average more than three grams. (The stated
amount used to be 6g, but it is generally agreed now this was too
high.) Most of us nowadays have far too much salt in our diet, and
there is no need to add salt to food for dietary reasons – we do it
merely for flavouring. Consuming too much salt increases risk of
heart disease and stroke.

Some individuals are recommended by their doctors to use
potassium chloride, sold under various brands as a salt alternative,
as it tastes salty, but does not contain the sodium ions that do the
damage with common salt. However, it is not recommended that
everyone do this as a matter of course, as potassium can have serious
side effects for some people, so such substitution should be done
only under medical guidance.

TV chefs (and most recipes) include far more salt than is necessary to perk up the taste buds – and, in fact, if you get used to it,
cooking without salt really isn't hugely different (because you can
pretty well guarantee that you will get most of the salt you need from
bread, cereals, processed foods etc. unless you live only off your
own produce). You will see a chef throwing in a 'pinch' of salt that is
more like a teaspoonful – far more than is needed in a single meal.

There are also foods that are high in salt, which are worth limiting. These include bacon, cheese, ham, pickles, smoked and salted fish or meat (including salami) and shop-bought gravy products, soy sauce, stock cubes and the like. Watch for salt content in things like crisps, bought sandwiches, sausages, table sauces and breakfast cereals, which can be high – though there is often a version with lower salt.

There is also a lot of confusion over the different types of salt. There are three ways to get hold of salt – to chemically produce sodium chloride, to mine it out of the ground (rock salt) and to extract it from seawater. (Most table salt is either mined or from seawater.) Our natural inclination is to think that mined salt is better than chemical, and sea salt better than mined, because somehow it is more 'natural'. But remember, this is a simple chemical. The only difference between the three is that mined salt will have a few more impurities, and sea salt a lot more impurities.

As far as saltiness and salt taste goes, there is no difference whatsoever between the three types. In blind tasting, no one can tell the difference between sea salt and rock salt. There is also no difference in your cooking between granulated salt and larger salt crystals. Some cooks think the large crystals taste different, because if you put them on your tongue, they will dissolve differently and have a different influence on your taste buds – but in your cooking, both dissolve, leaving no trace of the form in which they started.

By all means, use large salt crystals in a grinder on the dinner table, as they look pretty, but in your cooking you might as well stick to cheap table salt, which will be indistinguishable from the most expensive sea salt.

LINKS:
- Minerals in the diet – see **Vitamins and minerals** – page 111

Saturated fat

**Traditionally, saturated fat has been regarded
as the bad boy of diet, the guaranteed
artery clogger. But headlines have suggested
this is not true. What are the facts?**

If there is one thing dieticians and medics have agreed on for some time, it is that saturated fat is a bad thing, and specifically that it contributes to obesity and heart disease. But the facts are a little more complicated.

Fats are organic compounds that combine glycerol (also known as glycerine) and fatty acids, which are chemicals with a long chain of carbon atoms with a 'carboxyl' chunk on the end that makes them acids. The fat is saturated if those fatty acids are saturated, which just means that each of the carbon atoms in its chain has its full component of hydrogen atoms, rather than having a double connection to an adjacent carbon.

For a long time it has been recommended that we restrict the amount of saturated fats, found in dairy and fatty meats, in our diets as they were thought to increase risks of heart disease and push up cholesterol levels in the blood. However, in 2014 there was considerable coverage of a report suggesting that the link between saturated fats and heart disease was not proven. This is a useful report based on good science, but its implications are not that we should merrily eat as much saturated fat as we like.

In the research, 72 significant studies looking at the link between fatty acids and coronary disease were pulled together to get an overview. They found that there was no significant evidence that saturated fats increase the risk of heart disease. There is one major proviso – some of the studies were of people who already had risk factors (like smoking) for cardiovascular disease, or were already sufferers. This means that more research is necessary to be certain that these results apply to the population at large, particularly healthy people.

But it does indicate that it is likely that consumption of saturated fats is not in itself going to increase your risk of heart disease. However, there is no doubt that excess consumption of any fats can pile on the calories, which in turn does damage your heart. Keeping an eye on your total calories is still important – and given the proviso, it is probably best keeping to the current recommended daily limits of no more than 30g of saturated fat for a man and no more than 20g for a woman.

It is worth noting that the research *did* find a link between consumption of trans fats or hydrogenated fats (see separate entry) and heart disease, emphasising the importance of avoiding these.

LINKS:
- **Cholesterol** – page 28
- **Fat** – page 44
- **Trans fats** – page 109

Smoothies

Smoothies are often sold on the basis of being a great way to get some of our five a day. And they are very tasty. But they aren't a substitute for eating fruit.

For some reason that it's hard to pin down, eating fruit is a bit of a chore. Most of us can manage the odd banana, but eating an apple seems like positively hard work. Yet we know that most of us don't eat enough fruit, and a high proportion of us don't regularly make our five, let alone seven, a day. So what to do about it? Why not turn a whole pile of fruit into a delicious drink?

First, the good side. A smoothie usually counts as about two of your five a day and has lots of fruit-based goodness. Given that a smoothie tastes as good as or even better than many high-sugar fizzy drinks, surely we should be drinking more smoothies and less sugar water? Many of us struggle to get our kids to eat enough fruit – a smoothie seems an ideal way to get them off the fizzies and get more fruit into their system.

However, there are some problems here. It is true that smoothies can contain enough fruit to be the equivalent of two of your five a day, but they do two things to reduce that fruit's value. One is that they destroy any dietary fibre in the fruit, reducing the benefit for your gut. Secondly, they smash up the structures containing fruit sugars, making the sugar more likely to be absorbed. Drinking a smoothie gives you more sugar intake than eating the fruits that went into it. And that's important because there is a whole lot of sugar going on in a smoothie because of the volume of fruit contained in it – typically more than the same amount of a full-sugar fizzy drink (and obviously vastly more than a sugar-free fizzy drink).

We don't have to be totally prescriptive about sugar. It's fine to have some – it's just that most of us consume far too much. In terms of overall food value, a smoothie is significantly better for you than cola, say, with the same amount of sugar. So the occasional smoothie as a treat – perhaps once or twice a week – is not a bad thing … and does get you some of your five a day. But they should not be consumed daily as they simply contain too much sugar.

To put it into context, smoothies can have as much as 14g of sugar per 100ml, which means in a 250ml drink a total of 35g of sugar (1.5oz of sugar in an 8.5oz drink). The current UK recommended daily allowance for an adult for sugar is about 70g (2.5oz) for men and 50g (1.7oz) for women, but it has recently been suggested that half of this is a better target. This makes a single smoothie the whole of a man's maximum sensible intake and significantly more than the recommended maximum for women and children.

LINKS:
- **Five a day** – page 47
- **Fruit** – page 55

Spices

Herbs and spices have been used to make food more interesting ever since we went beyond eating food straight from being collected. But have they any dietary benefits?

Herbal medicine is covered separately in the health section, but here we are considering the impact of herbs and spices on our diet. These add-ons to cookery largely contribute flavour and colour, though sometimes they are also used as a preservative.

There tends to be very little nutritional value to them as they are largely used in very small quantities, though some spices are good contributors to mineral and vitamin requirements if quantities from a teaspoon to a dessertspoon per portion are added. (This is around 15–30 grams or 0.5–1 ounce, though it will vary from spice to spice. The point is that it is far more than is usually used.)

Many herbs and spices have an antioxidant component, but as described elsewhere, there is no proven benefit from having antioxidants in the diet. Similarly, you might see an article telling you that curcumin, the active ingredient in the spice turmeric, could slow down the onset of dementia. Unfortunately, all that this claim is based on is that some genetically modified fruit flies had improved lifespans when exposed to curcumin (normal fruit flies actually died sooner). It tells us nothing about its impact on humans.

Technically, a lot of our favourite strong spices are poisonous, though as with all poisons, taken in small enough quantities – as is usually the case with, say, chilli – they do not cause any problems. Capsaicin is the alkaloid poison that gives chillies their kick. It's not in large enough quantities in all but the most vicious chillies for it to be a danger, but it is banned as a food additive, where it comes in a form sixteen times stronger than the most powerful plant.

When we taste a chilli, the capsaicin fits on to receptors in the tongue and skin that usually respond to a heat burn, making it feel as if we are actually burning, though in fact this is just the chemical interacting with your inbuilt sensors. Even though we aren't taking enough to cause serious problems, some people with gastrointestinal problems can find that chilli triggers reflux and stomach pains.

So, don't expect a lot of nutritional benefits from your herbs and spices – even though you will inevitably find plenty of websites extolling their antioxidant and other powers – but you can't beat the way they can improve flavours.

LINKS:
- **Antioxidants** – page 7
- **Herbal medicine** – page 293

Spreads

More people use spreads than butter these days, but what's the best choice as far as diet is concerned?

It is worth noting straight away that the term 'spread' is not just a modern word for 'margarine'. Margarine was the first of the non-butter spreads and, to be honest, it was pretty unappetising. Napoleon III of France had offered a prize for a cheap alternative to butter, aimed at the poor and the army. The winning product, 'oleomargarine', was a nasty product of beef fat with some added vegetable oils and water. This had to be tinted yellow as it was, by nature, an unpleasant shade of off-white.

Over time, the tallow content was phased out, to be replaced by vegetable oils, but that gave a consistency problem that was addressed using the technique that gave margarines a bad name. By using semi-hydrogenated vegetable oil, producers could get an appropriate texture, but at the cost of incorporating trans fats, which proved to be bad for health.

Modern spreads in the UK contain no trans fats and are made primarily from vegetable oils and water. Broadly, they divide into four categories: butter-based spreads, oil-based spreads, low-fat spreads and cholesterol-reducing spreads.

The butter-based spreads attempt to combine the taste of butter with spreadability, usually by reducing the fat content from the butter by using more buttermilk, and by mixing in some vegetable oils. The pure vegetable oils tend to be primarily either sunflower or olive oil-based, although frequently the headline oil, like olive oil, is not the only oil contained. Low-fat spreads have extra water, reducing the fat content from the typical 80 per cent of a full-fat spread in exchange for arguably making them slightly less effective as spreads and unsuitable for frying. And then there are the cholesterol-reducing spreads.

These are usually a sunflower oil spread with added plant sterol and stanol esters. These are effectively very low dose variants of the statin medicines used to reduce cholesterol and do have a genuine effect in lowering 'bad' cholesterol levels, though the percentage change is relatively small – around 10 per cent reduction if consumed daily – so any serious elevation of cholesterol is better treated with medication.

Until recently we would have said that the polyunsaturated fats in vegetable oil spreads were better for the cardiovascular system than the saturated fats in butter-based spreads, but this has been pretty much discounted, so the choice is primarily down to one of taste, though the fact that we generally eat far more fat than we need for a balanced diet suggests that it is still advantageous to go for one of the low-fat alternatives available in either type.

There has been a tendency to assume that olive oil is somehow healthier than sunflower oil, in part because it is a significant component of the Mediterranean diet, and the people on that diet tend to do better in terms of heart problems. However, there is no evidence that the olive oil plays a contributory factor, so again it is primarily a matter of taste.

LINKS:
- **Cholesterol** – page 28
- **Trans fats** – page 109

Starch

**Our staple foods have large quantities of
starch, a common carbohydrate. How should
starchy foods be represented in our diet?**

If you are a certain age, you probably associate starch with the substance that made clothes uncomfortably stiff after washing, but it is a widely available carbohydrate, a long compound that chains together glucose molecules and that is used as an energy store by plants.

Starchy foods make up our staples – bread (via wheat, etc.), potatoes, rice and pasta (wheat, etc. again). The best advice seems to be to make up around one third of your diet with starchy foods, making them, as much as possible, the high-fibre versions like wholemeal/wholegrain bread/pasta, brown rice and potatoes with their skins on.

LINKS:
* **Carbohydrates** – page 19

Storage of food

**Our ability to eat produce from around the world
at any time of year is unparalleled in history, but
that does mean some of the 'fresh' food we buy has
been frozen or in cold store for some time. Does that
make any difference to the dietary value of the food?**

Walk around a supermarket and you will find an abundance of fresh and frozen foods. Inevitably, most of it will have been stored for a period of time, in transit from growing locations around the world and in the retail process. But what effect does this have on the nutritional value of the food?

Strangely, we can usually be more certain about the pedigree of frozen food than fresh. Once frozen, and as long as it is within the date range, the freezer food will generally hang on to pretty well all of its nutritional value. What's more, it is usually frozen very soon after picking, maximising the ability to hang on to the goodness. The only proviso is that some frozen fruit and vegetables are dipped in boiling water (blanched) before freezing to ensure that yeasts, etc. are inactive – this can reduce vitamin C levels by up to 20 per cent.

By comparison, fresh food will lose some nutritional value during the period that it is stored. Soft fruits and green vegetables can lose around 15 per cent of their vitamin C content per day, for instance, when stored at room temperatures, and other vitamins and minerals will be leached out over time. Even when chilled there is loss. In one week at 4°C (refrigerator temperature), peas will lose about 15 per cent of their vitamin C, while green beans lose a massive 75 per cent. Similarly, the A, B and E vitamins are sensitive to heat and light, and depending on the transport methods can be lost in considerable quantities from foods undergoing long journeys in hot countries.

This means that even farmers' market-fresh produce, which may well only spend a couple of days between leaving the ground and reaching your shopping bag, will be slightly nutritionally worse than a frozen equivalent, while the fresh produce in a supermarket will almost certainly have been stored for several more days, though it is more likely to have been chilled, which will reduce the rate of loss of nutritional value. When transported by ship, fruit and vegetables (though inevitably refrigerated) could well go several weeks from picking to eating.

Another problem with a lot of the fresh food that we buy from abroad is that it tends to be picked before it is ripe, allowing for the fact that it will ripen during the time it is being shipped. But this means that the fruit or vegetable is not at its nutritional peak when harvested, and it won't gain extra nutritional value once picked. This is a strong argument for going as much as possible for seasonal, locally-grown fruit and vegetables, which can be picked when ripe and still arrive with the consumer while edible. (Frozen food does not suffer from this problem, as it is picked when ripe.)

We have always kept some fruit and vegetables for long periods to cope with seasonal variations. Potatoes and apples, for instance, have traditionally been stored, and in the modern equivalent, apples and pears can be kept in low oxygen, high carbon dioxide stores for up to twelve months with relatively little deterioration in nutritional value – though this is a relative rarity.

There is no doubt, though, that fresh food can look and taste better than frozen, and buying locally, ideally from the farmer, is the best way to maximise the nutritional value of food by minimising the 'ground to plate' time. (Or, better still, grow your own!) As the organic section explains, buying organic provides no extra nutritional value, and it is perfectly possible for food that is labelled organic to have had exactly the same delays in shipping around the world as any other food. Until supermarkets give information on when a crop was picked, there will always be uncertainty on how long foods have been in transit and storage, with inevitable loss of nutritional quality.

LINKS:
- **Organic food** – page 74

Sugar

We love sugar, but it has limited food value and is definitely not good for us in any significant quantity. On average we probably eat about three times the amount we should.

Sweetness is a universal desire in humans and some animals. (Dogs, for instance, love sweet things, but cats don't, as they have no sweet sensors on their tongues.) Yet sweet things don't do us much good, and sugar in particular is one of the worst aspects of the modern diet.

You might wonder why we universally crave something so bad for us (and the same goes for fat and salt – the reason why 'junk food' is so appealing). It's because our programming lags far behind our environment. Over time we've gone from a situation where it was important to cram in the calories because we didn't know where our next meal was coming from, to one where food is all too easy to get hold of. We are programmed for scarcity in an environment of plenty – which means our minds have to work overtime to suppress the (usually dominant) demands of our bodies. All too often, the tongue wins over the head.

Sugar is a source of energy, but isn't a necessary part of the diet – there are better ways to get that energy. Meanwhile, there is no doubt whatsoever that excess consumption of sugar results in an increase in diabetes and obesity, and can also have a terrible impact on the teeth. And all we get in exchange is a nice taste. New UK recommendations suggest a daily allowance for an adult for sugar of no more than 35g for a man and 25g for women (1.2 and 0.8oz respectively), but this is extremely difficult to achieve.

As we see in the **Smoothies** entry, a single smoothie can push us over that limit. And the problem is that plenty of the processed foods we consume have sugar added to make them more attractive. A 2014 survey by Action on Sugar found some remarkable figures:

Product	Size	Grams of sugar
Starbucks Caramel Frappuccino with whipped cream (with skimmed milk)	Tall (small)	44.3
Coca Cola Original	330ml	35
Pepsi Regular Cola	330ml	35
Mars chocolate bar	51g	30.4
Pret A Manger Very Berry Latte (with milk)	295g	26.9
Müller Crunch Corner Strawberry Shortcake Yogurt	135g	23.6
Sharwood's Sweet & Sour Chicken with Rice	375g	22.1
Cadbury Hot Drinking Chocolate (with semi-skimmed milk)	200ml	22.1
Yeo Valley Family Farm 0% Fat Vanilla Yogurt	150g	20.9
Solero Exotic Ice Cream	88ml	17
Kellogg's Frosties (with semi-skimmed milk)	30g	17
Butterkist Toffee Popcorn	25g	16.5
Glaceau Vitamin Water, Defence	500ml	15
Heinz Classic Tomato Soup	300g	14.9
Ragu Tomato & Basil Pasta Sauce	200g	13.8
Kellogg's Nutri-Grain Crunchy Oat Granola Cinnamon Bars	40g	9
Pot Noodle Curry King Pot	114g	7.6
Heinz Tomato Ketchup	15ml	4
Heinz Salad Cream	15ml	2.6
Hovis Soft White Bread Medium	40g	1.4

With numbers like that, it is very easy to exceed the recommended daily amount.

What can we do about it? Avoid the really high-sugar drinks – switch to a low-calorie or zero-calorie alternative. A zero-calorie cola, for instance, may initially taste foul, but we do get used to them and begin to enjoy them, without that sugar hit. Check the sugar content when you eat processed foods – watch out for sugar as you might for fat content. Sugar can creep in where you might not expect it. For instance, skimmed milk has very little fat indeed, but a 250ml glass of milk contains 12.5g of sugar – about half the new recommended daily maximum for a woman.

LINKS:

- **Artificial sweeteners** – page 9
- **Five a day** – page 47
- **No added sugar** – page 68
- **Smoothies** – page 93

Superfruits

Most of the benefits of superfruits (and other superfoods) are unproven or dubious. There's nothing wrong with enjoying them as part of your diet, but don't expect them to have extra health benefits.

Quite a few foods, especially fruits like pomegranate and blueberry, are hyped as having health benefits, whether it's as a natural cure for various ills or just a way to help your skin look better. These claims are made so frequently that the superfruits also turn up in a whole range of processed products from juices and cereals to makeup and shampoo.

As far as the food industry is concerned, the use of the term 'superfruit' (like the term 'organic') is simply a way to charge a higher margin on a product. But this doesn't in itself mean that beneficial effects are exaggerated. However, all the evidence is that these fruits have very limited value over and above any other fruit that doesn't have that extra profit margin added.

What are these wonder fruits claimed to do? The term we used to hear most often was 'antioxidants', which are usually present in relatively high quantities in chemicals like bioflavonoids and polyphenols. As described in more detail in the **Antioxidant** section, these are valuable within the body to protect it from free radicals, which can cause cancer and heart disease – but there is no evidence that we benefit from *consuming* additional antioxidants, as in superfruits, and there is evidence that some antioxidant supplementation can actually *cause* cancer. If you really want to up your antioxidant intake, broccoli is a better source than superfruits, as it provides a much wider range of beneficial chemicals, and hasn't got the high sugar load.

Superfruits certainly also do give a good dose of vitamin C, but a deficiency of this vitamin is rare in developed countries, as lots of foods contain it. Just the skin of a potato will give you nearly half the daily-recommended amount. There is no good evidence that overdosing on vitamin C has any health benefits.

Like all fruits, superfruits contain sugars. Be particularly wary of drinks. Smoothies really pack in the sugar, while many superfruit drinks have to be sweetened because they are often made of the more bitter or astringent fruits. This sweetening may well be done with apple or grape juice or other fruits to be able to use the 'no added sweetener' label, but fruit sugar is just as bad for you as any other. Some superfruit drinks have as much sugar (five teaspoons in 250ml) as a cola.

You may be surprised at this point if you have heard of studies that 'prove' that, for instance, eating pomegranates increases your life expectancy. Unfortunately, such claims have often been made and yet there turns out to be no such study. It's easy to say one exists – far more difficult to find one in the scientific literature, because they aren't there.

Don't see superfruits as a health benefit, then. You are still better putting, say, blueberries, as a sweetener on your cereal than sugar alone, as you will be adding fibre and other benefits. But fruit is fruit.

LINKS:
- **Antioxidants** – page 7
- **Chemicals** – page 21
- **Vitamin C** – page 114

Supplements

If you are missing essential nutrients from your diet, supplements of vitamins and minerals are important. But for most healthy people they provide no measurable benefit.

With a good, broad diet we are unlikely to need supplements. Some do – people with some medical disorders or who have a restricted diet, whether vegans or with poor access to a broad range of food, may need supplementation. But for the rest of us, there is no known benefit.

This is useful to know, as we spend around £750 million a year in the UK alone on vitamin supplements. The fact is that you will get plenty of vitamins and nutrients from that normal diet, even if it is unhealthy in terms of fat or sugar levels.

Equally, it is worth being aware that there is no known benefit to health (for instance to the 'immune system' as the adverts often claim) from taking vitamins and minerals as supplements and pushing up the levels above those provided by a normal diet. A detailed review of studies has been performed on the effects of vitamin supplements on health and overall there was no significant benefit.

Most of the supplementary substances are perfectly safe to overdose on, as they just pass through the body unused, but it is important to be careful with vitamin A (see the **Vitamins and minerals** section) if, for instance, you eat liver and take vitamin A supplements you could reach a dangerous level.

Large-scale studies of antioxidant supplements, particularly beta-carotene, suggest that they even *increase* the risk of dying by a small but significant amount. Quite why they are still available for sale is incomprehensible.

LINKS:

- Antioxidant supplements – see **Antioxidants** – page 7
- Supplements to enhance cognition – see **Brain food** – page 158
- **Vitamins and minerals** – page 111

T

· · · · · · · · ·

Tannins

**Tannins are chemicals found in tea, red wine, fruits
and juices. Though some are powerful antioxidants,
there is no known health benefit from them.**

We think of tannins as the substances in tea and red wine that make
them astringent – they give us that lip-puckering effect when we
drink them. The name comes from the same source as 'tan' and
'tannery', originally referring to the bark of an oak tree, which was
one of the first substances used in tanning leather. (So when you get
a tan, your skin is either becoming leather-like or like tree bark.)

That sharp taste suggests that tannins may have developed as a
natural insecticide, though they also have a role in regulating plant
growth and decomposition. Their job is to bind on to proteins and
precipitate them, changing the structure of the dead material, which is
why they stiffen and waterproof animal skin as they turn it into leather.

Tannins seem to be neutral as foodstuffs. They have no food
value, and have no identified health benefit (though it is just possible
they are involved in the apparent health benefits of red wine). But
equally they are not bad for us, and many of us enjoy that peculiar
effect when a tannin hits the tastebuds. It is perhaps the same per-
verse enjoyment as we get from the aggressive chemical capsaicin
found in chillies. We like to push our senses just outside their comfort zone, and tannins fit the bill.

LINKS:
- **Antioxidants** – page 7
- **Red wine** – page 87
- **Sunburn** – page 320

Trans fats

Of all the dietary fats, the ones that everyone seems to agree have no redeeming features are trans fats. Avoid!

Fats are organic compounds that combine glycerol (also known as glycerine) and fatty acids, which are chemicals with a long chain of carbon atoms with a 'carboxyl' chunk on the end that makes them acids. The fat is unsaturated if one of those fatty acids is unsaturated, which just means that at least one of the links between carbon atoms in its chain is double, rather than the usual single bond.

Natural unsaturated fats generally have a kink at the double bond – these are known as 'cis fats'. But some, mostly artificially created, unsaturated fats have a straight double bond. These 'trans fats' tend to occur as an accidental side effect when natural unsaturated fats are hydrogenated, a process that makes them harder and more suitable for baking. Hydrogenation is also used in processed foods like biscuits and cakes to extend the shelf life, and in some frying products.

There have been doubts about trans fats for a long time, and a major 2014 piece of research, covering 72 existing studies, showed that there is a significant association between the consumption of trans fats and an increased risk of heart disease. However, most UK consumers don't eat a lot of trans fats (they are much more widely used in the US, for instance), tending to average only about half the recommended maximum – but they are still worth cutting down.

It is entirely possible to avoid trans fats by cutting out products like biscuits and cakes, ready meals and fried food where you don't know what it has been fried in. There are some natural trans fats in meat and dairy, but at relatively low levels that don't need to cause concern. You can also check by looking for mentions of (partially) hydrogenated fat in the food labels. Equally, however, supermarkets are increasingly responding to the demand to avoid trans fats. Some have cut them out entirely and others are on the way to doing this. Check with your supermarket.

Like almost all research, there is a proviso. The effect on the risk of heart disease was discovered only as a connection with consumption of trans fats, rather than with levels of trans fats in the blood, which did not show a link. In such a case there could be danger of 'correlation without causality', typically because a third factor is causing both the increase in trans fat consumption and the increase in heart disease.

What could be the case, for instance, is that people who take less care of their bodies are more likely to eat foods containing trans fats, like biscuits, fried food and ready meals, and are more likely to avoid exercise and be overweight. It could be incidental that the trans fats are present. But given that there is no need whatsoever to eat trans fats, and some supermarkets are able to totally remove trans fats from their product lines, there is every reason to take the precautionary approach and avoid trans fats in your diet.

LINKS:
• **Fat** – page 44

V

• • • • • • • •

Vitamins and minerals

Vitamins and minerals, also known as micronutrients, are essential for good health. Most of us get all we need from a balanced diet plus sunlight. Here's a quick guide to some of the essentials.

Vitamin A – important during growth and supports the eyesight. A deficiency may result in night blindness. Provided in the diet by carrots, green leafy vegetables, egg yolks, cod liver oil and liver. Some margarines are enriched. An excess is harmful, which is why it's recommended you don't eat liver more than once a week (one portion contains over 30 times the recommended daily amount).

Vitamin B1 (thiamine) – supports the nervous system and heart. A deficiency will result in loss of concentration, confusion and exhaustion. At the extreme, the deficiency is known as beri-beri. Present in rice, milk, peanuts, yeast and yeast products (like Marmite and Vegemite), wholemeal bread, cereals, liver and pork. Some breakfast cereals are enriched. Avoid having alcohol and coffee with all meals as they destroy B1 (as do high temperatures).

Vitamin B2 (riboflavin) – needed for the eyesight, skin, nails and hair. A deficiency results in itchy skin and eyes. Present in cheese, milk, green leafy vegetables, fish and liver. Some breakfast cereals are enriched. Alcohol and bright lights destroy it.

Vitamin B3 (niacin) – supports the nervous system and required for good skin. Deficiencies can result in weakness, loss of appetite, dermatitis, diarrhoea and, at the extreme, dementia. Found in wholegrain products, peanuts, sesame seeds, fish and most meat. Some breakfast cereals are enriched.

Vitamin B6 (pyridoxine) – like B3, required by the nervous system and for good skin. A deficiency can result in sleeplessness, irritability, skin problems and, at the extreme, convulsions. Found in beans, bananas, fish, chicken and pork. Can be destroyed by alcohol, over-cooking and oestrogen.

Vitamin B12 (cobalmin) – required for the formation of red blood cells and nerves. A deficiency can result in anaemia and damage to the nervous system. Mostly sourced from dairy, fish and meat, so vegans may require supplements, though short-term abstinence is not a problem as the liver can store up B12 for up to five years. Destroyed by alkaline substances like baking powder.

Vitamin C – see separate section.

Vitamin D – see separate section.

Vitamin E – an important antioxidant, removing free radicals. A deficiency can result in physical weakness and infertility. Found in leafy green vegetables, broccoli, soya, vegetable oils, nuts and eggs. Can be destroyed by heat, cold, chlorine and oxygen.

Vitamin K – helps produce special blood factors that allow clotting and reduce haemorrhage, as well as supporting bone and kidney functions. Around half our requirement is produced internally by bacteria in the gut. The remainder is obtained from leafy green vegetables, wheat and some meats.

Folic acid (folacin) – helps in the production of red blood cells and essential in the first trimester of pregnancy. A deficiency can cause anaemia and birth defects. Naturally occurs in carrots, beans, leafy green vegetables, whole wheat, yeast products and liver. Frequently added to breakfast cereals. Destroyed by heat, oxygen, sunlight and oestrogen.

Minerals – a number of minerals are essential in small quantities, including calcium, iron, magnesium and zinc. Our main source of calcium is dairy, so vegans should take supplements (soya products are often enriched to provide this). Iron is found in a fair number of fruits and vegetables, but it is easier to absorb from meats. A deficiency results in anaemia. Magnesium is hard to miss as it is in most foods and we don't need a lot. Usually a deficiency is caused by prolonged vomiting, diarrhoea or alcohol consumption. Zinc is most easily obtained from meat and eggs – it is difficult to absorb from vegetables, particularly whole grains, so vegetarians may need supplements. A deficiency leads to delayed healing and skin irritation.

Salt – see separate section.

LINKS:
- **Dairy** – page 32
- **Salt** – page 90
- **Supplements** – page 106
- **Vitamin C** – page 114
- **Vitamin D** – page 115

Vitamin C

Like most vitamins, we get plenty of vitamin C from a balanced diet, though it doesn't do any harm to boost it a little. There is no evidence it helps with colds or cancer, however.

There is probably more nonsense talked about vitamin C than any other micronutrient. There is no doubt that we need it. A deficiency results in bleeding gums, tiredness and slow healing wounds, with the crippling disease scurvy at the extreme. Natural sources include citrus fruits, berries (especially kiwi), tomatoes, potatoes, peppers and leafy green vegetables. You do get vitamin C from cooked food, but raw is best, preserving around 25 per cent more of the vitamin. A glass of orange juice or a portion of strawberries give you a full dose – even a portion of potatoes provides around 15 per cent of requirements.

There have been numerous claims that massive doses of vitamin C – typically 20 or 30 times the recommend daily amount – can cure colds or even have a therapeutic effect on some cancers. Doses at or above these levels (anything above one gram a day) should only be taken under medical supervision, as there are potential dangers. But more significantly, there is no reason for doing this, as there is *no* good scientific evidence to back up taking these massive vitamin C doses. They do not reduce your susceptibility to colds or cure cancer.

One of the reasons this treatment has more credibility than most odd suggestions is that it was supported by the Nobel Prize-winning scientist Linus Pauling. Pauling was a chemist and won his Nobel Prize for his explanation of chemical bonding – the way atoms link together. He had no medical expertise, and there is no reason that having a Nobel Prize in one subject gives any authority in another.

LINKS:

• **Vitamins and minerals** – page 111

Vitamin D

The good news about vitamin D is that it only takes a few minutes of sunlight to get a top-up.

We have got so used to warnings about the dangers of the sun that it's easy to forget that a little bit of sunlight on the skin is a very good thing. Vitamin D is essential for strong bones and teeth, while a shortage can lead to rickets and an increased risk of some cancers and of osteoporosis in older people.

To keep your vitamin D levels healthy requires a regular exposure to direct sunlight (not through glass). And for the relatively short period of time you are exposed, it's important not to be covered in high factor sunscreen. Even sunscreen as low as factor 8 cuts vitamin D production by over 97 per cent.

A small, controlled amount of summer sun without sunscreen is a good idea for vitamin D production. The exact approach to take depends on your location. In the latitudes above 37 degrees north (which is most of Europe and North America), recommendations for those with fair skins are to have exposure around the middle of the day, starting around two minutes per side and working up to a total of around fifteen minutes a day, exposing up to half the body. These amounts need to be varied depending on your skin type and location. A dark African skin tone reduces production of vitamin D by 80–90 per cent over fair Celtic. In countries with stronger sunlight, like Australia, just a few minutes of exposure is sufficient. Check your local meteorological bureau for more detailed guidance.

What is essential is that you should cover up or get in the shade if you begin to feel hot or uncomfortable, or if your skin gets a slight pinkness, and that outside this window of vitamin D sunbathing, you should stay in the shade or use sunscreen of SPF 30 or above. It's also best, apart from this exposure, to stay out of the sun in the four hours around the middle of the day. In hotter climates, move the time further away from the middle of the day. For children, or if in any doubt, take medical advice.

Although it has to be handled carefully, sun exposure is by far the best way to get vitamin D. Some foods like oily fish contain vitamin D and in some countries (Canada and the US, for instance) milk and margarine are fortified with the vitamin, but it isn't easy to absorb from foodstuffs. A lot of older people take vitamin D supplements to try to stave off osteoporosis, but a New Zealand study published in the *Lancet* in 2013 found that vitamin D pills do not improve bone density in healthy adults.

LINKS:
- **Sunburn** – page 320
- **Vitamins and minerals** – page 111

W

. . . .

Weight loss foods

**We are always looking for quick and easy solutions
to lose weight, so it's tempting to fall for foods
that are claimed to help trim the pounds.**

If you receive any quantities of email spam, you will probably have
had offers for weight loss products alongside the attempts to break
into your bank account or to sell you medication to help your sex
life. Notably, green coffee beans and raspberry ketones keep popping
up with painful regularity, if my inbox is anything to go by. But can
they really help you shed that unwanted bulge?

Green coffee beans are standard coffee beans before roasting.
As you'll discover elsewhere, coffee certainly has a range of impacts
on the human body, but it seems unlikely these include weight loss.
This might seem strange when those emails always mention that this
is proven by a study. And there is one. But it is only a single study
with sixteen participants, which did not have good controls and
doesn't tell us anything of statistical significance. Add in that the
study wasn't registered, wasn't carried out by appropriately qualified
experts and was incorrectly constructed, and there is effectively no
evidence at all.

As for raspberry ketones, these are the intense chemicals that
give raspberries their distinctive smell. So intense, in fact, that they
are only present in tiny quantities in the fruit, so have to be made
artificially to be used in a diet pill. (Not that there's anything wrong
with that, but don't assume from the name that this is a natural
product derived from raspberries.) According to the advertising:
'Raspberry Ketone increases the hormone adiponectin, which
burns away fat cells and lowers insulin levels. And best of all, it is

all-natural and clinically proven.' We are told that: 'You can lose up to 20lbs in just 2 weeks!'

Sadly there is no evidence to back up the claims here. There have been two studies showing weight loss in rats with a diet consisting of around 2 per cent raspberry ketones – but other studies have contradicted this, showing no weight loss. This is hundreds of times more of the ketone than is present in weight loss pills, and even if it did produce an effect in rats (which the other studies did not find), it does not mean that there would be an effect in humans. There have been no trials on humans whatsoever. Nor has there been any testing to determine what levels are safe. At the time of writing, raspberry ketones are considered 'novel foods' that have not been suitably tested and are not legal for sale in the UK. (This is also true of another weight loss supplement, *Acacia rigidula* or Blackbush acacia.)

Short of medical intervention, there is only one simple way to lose weight. Eat less and exercise more. Use more calories than you consume and you will lose weight. Of course, we want to sit on the couch all day and eat whatever we like, then just make it go away. But life isn't like that.

LINKS:
- **Coffee** – page 29
- Coffee and the brain – see **Brain food** – page 158

EXERCISE

. . . .

Whatever all the diet and fitness books tell you, science is very clear that there are just two basic requirements to maximise your health – a good balanced diet and appropriate exercise.

The exercise part is not about entering a marathon every day, nor does it require expensive gym membership. It's just a matter of having a reasonable level of exercise – say twenty minutes of fast walking or jogging – at least five days a week. Of course, there are all sorts of exercise you can take to perform specific tasks, like building up muscles. But in terms of basic health, it doesn't have to be any more complicated.

Just like in the dietary field, there is a lot of myth and nonsense built up around exercise. One obvious example is the whole business of warm-ups. For many years, those about to undertake strenuous exercise were told they needed to do various warm-up exercises, including static stretching – the sort of thing where a runner, for instance, will push his or her leg into a stressed position and hold it for some time. This particular type of warm-up has actually been proved to be a disadvantage – yet you still see people doing it all the time. The same goes for many post-exercise cool-downs. At the time of writing, the generally excellent NHS Choices website still recommends a whole routine of static stretches after exercise, now shown to have negative effects on future performance.

The main problem with exercise is often motivation. It's too much trouble to go for a run. It's raining today, so you won't bother to go out. Life's too short. And these restrictions are totally understandable. So, one essential when looking at how science can improve our health through exercise is to take some hints from psychology on ways to encourage ourselves to undertake exercise.

The problem here is that our brains are very much orientated to short-term gratification. When you live a hunter-gatherer life, you grab what you can while it is available, as you never know what things will be like in the future. Our lives are very different now, but our brains have not caught up. This is why politicians find it so difficult to commit to long-term policies, and why we all find it easier to eat a doughnut than stick to a diet – or to sit in front of the TV rather than exercise.

What's happening is that the benefits we are putting on offer to ourselves – better health, living longer – are all long-term, distant goals. We are asking ourselves to suffer in the short term to benefit in the long term, and our brains really struggle with this. So, the best way to ensure you get your exercise in is to look for short-term gratification and to build it into activities that you would ordinarily undertake anyway. Don't drive to the shop if it's nearby; walk. Give yourself short-term rewards (ideally not a doughnut) to encourage yourself to undertake exercise. It's a case of using your modern, canny ideas to overcome the old, short-sighted wiring in your brain.

A

· · · · · · · ·

Aerobic exercise

**One of the problems with exercise is the tendency
to use terminology that doesn't mean a lot to us. So
what is aerobic exercise, and why is it a good idea?**

We often hear about 'aerobic' exercise, which sounds like exercise in
fresh air, but what is it all about? The terms 'aerobic' and 'anaerobic'
mean 'oxygen-based' and 'not requiring oxygen' respectively. They
are used in biology to divide bacteria into two kinds – those that,
like us, need oxygen to live and those that can thrive in environments where no oxygen reaches them.

In a sense, all exercise is aerobic, because we use oxygen in the
process. But in the world of fitness training, 'aerobic exercise', sometimes referred to as 'cardio', is used to mean the kind of exercise that
is sustained and of sufficient intensity to raise the heart rate without
being an all-out push. It uses some calories, which are burned in
an aerobic fashion. So we're talking about exercise like fast walking,
jogging, paced running, cycling, rowing and swimming.

Technically speaking, you don't even have to get the heart rate
up to perform aerobic exercise. Any movement – even standing still
compared with sitting down – uses some calories. But for health
purposes, the aerobic exercise should be sufficient to get the heart
rate up by a noticeable amount. If you don't feel that you are pushing
yourself a little there will be relatively small health benefits over and
above reducing your build-up of calories.

The term 'aerobic' is used to contrast with 'anaerobic exercise'.
Anaerobic exercise refers to intense bursts of exercise, typically
lasting no longer than two minutes, and required for sports like
sprinting and for muscle building. Although technically anaerobic

exercise isn't truly anaerobic, it does put sufficient load on the body to require energy to be taken from bodily stores rather than the usual aerobic process that combines the oxygen we breathe with chemicals from our food to produce energy.

Anaerobic exercise produces lactic acid, which was, for a long time, considered a serious problem, and one of the reasons for complex cool-down procedures after strenuous exercise, particularly massage. But modern research suggests that, except in the most extreme cases, lactic acid build-up is not an issue that requires any action to be taken – and after significant studies it seems that massage actually reduces blood flow and prevents lactic acid being carried away, so it is certainly not the way to reduce build-up.

LINKS:

- Aerobic exercise and mental function – see **Exercise and the brain** – page 169

Afterburn

It's not unusual to find a diet/exercise book claiming that you continue to burn calories after exercise. Unfortunately there is no basis for this claim.

This was a new one on me, but if you trawl the diet guides and fitness books you are likely to find a phenomenon called 'afterburn'. The idea is that as a result of exercise, your metabolism is speeded up, and because of that, after the exercise you will burn up more fat calories, even if you are just sitting around. It's an appealing thought and has a kind of common sense feel about it. Surely exercise does speed up the metabolism and increase your calorie consumption afterwards?

Unfortunately, common sense and science don't often go hand in hand. The idea of our metabolism being something like an engine that takes time to slow down after it has been 'revved up' is a nice mental image, but it seems that the diet and fitness guidance was based on guesswork rather than real data.

When a trial was made using sophisticated equipment that measures exactly how many calories are burned, and how much these come from fat, the results were conclusive. Not only was no extra body fat consumed after exercise, if anything there was a slight dip in fat burning over the whole day, as during the exercise there would be an increase in carbohydrate burn instead.

It might seem mean, but the balance remains simple. Burn calories when active, reduce to a tick-over rate when inactive. There is no afterburn.

LINKS:
• **Calorie intake** – page 17

B

· · · · · · · ·

Breathing and relaxation

**Approached in the right way, breathing and
relaxation can be a great way to have an exercise
that helps with calmness and reduces stress.**

It might seem that 'breathing and relaxation' would be a very short
entry. You need to breathe. You need to relax. Do it. But there's
more than one way to breathe, and relaxation covers a whole range
of possibilities.

Breathing the right way can help you unwind and counter stress.
First, you need to identify how you breathe. Broadly, there are two
approaches – using the chest muscles, where the breath responds
to a rise and fall of the chest, and from the diaphragm just above
the stomach.

Breathing with the diaphragm is more controlled (good sing-
ers use this) and gives a more effective breath. Try giving it a go
first. Stand upright but not tensed (shake your shoulders if you feel
tense). Take a deep breath, feeling your chest rise. Now try keeping
your chest in the 'up' position as you breathe in and out. You should
feel a tensing and relaxing of the diaphragm just above the stomach.

Once you've got the hang of breathing this way, lie down or sit
comfortably. Close your eyes. Breathe regularly, mentally counting
to five as you breathe in, hold it for a count of two, then breathe out
for five again. Try to keep the breathing action to the diaphragm.
You should feel tension reducing – and the nice thing about a
breathing exercise is you can do it (minus the closed eyes) in pretty
well any situation.

Now let's extend the relaxation. You are going to focus on different parts of your body, starting from the head and working down

to the toes. As you consider each part, tense the muscles, hold them for a count of two, then relax them slowly with a long breath. Try to keep your thoughts on the breathing and the muscles. Don't think of anything else.

Finally, relax as much as you can and try to keep your mind as empty as possible for a minute. When you have finished, move slowly to begin with – don't jump up and start bouncing around immediately.

You can extend this process as meditation, something that doesn't have to be all New Age and mystical.

Broom and mop efficiency

It seems a strange thing to put in the 'exercise' section – but there certainly is exercise involved in using a broom and a mop, and though mops and brooms haven't had much attention from science, they do provide some quick lessons.

As far as sweeping the floor goes, a 'push broom' – the kind with a brush head on the end of a broomstick – is significantly more efficient than a traditional 'twig' broom. It takes less work to move the same amount of dust. In terms of technique, the push action is best to move the largest amount in a single stroke, while the pull action gathers more of a relatively small amount of debris. This suggests the best strategy is to use the push to clear large materials first, the pull on any areas that still require sweeping.

As far as mopping goes, there has been a considerable amount of effort put into finding the most efficient mopping action by commercial cleaners who want the best clean in the shortest time. They have found that, rather than use a simple broom-like front and back motion, as most of us tend to by default, by far the best approach is to cover a strip of floor side-to-side using a snake-like motion. Of course, this should be done backwards, so you don't walk on the floor that you have already cleaned. This technique assumes that you are using a multi-stranded mop (rather than a flat 'squeegee' mop) which can be used with the snake action.

C

• • • • • • • •

Cycling

There seems little doubt that riding a bike is one of the best forms of exercise, although it does have some disadvantages when compared with fast walking or jogging.

The Lycra-clad bike rider is the epitome of the health and fitness enthusiast, and there is no doubt that riding a bicycle, whether you are a devotee with all the gear or a casual, fun cyclist is one of the simplest and most straightforward means to improve your exercise routine. It has the significant advantage over walking or jogging that it can be built in to a longer journey like a commute to work, where the walker is limited to a practical range of about four miles.

However, there are some negatives. Riding a bike is a less energetic way to cover a distance – you will burn up more calories walking or jogging and get better aerobic exercise over the same distance with a fast walk or a jog. (The distinction is less clear on hilly territory, where there is the extra effort of getting the bike up the slope, but it still holds.) So, for distances under four miles it is better not to use the bike. What's more, the bike involves a lot more expensive equipment, it has to be stored at your destination, and there's the non-trivial matter of safety.

Most people who ride regularly on the road have stories of near misses and potential disasters when they encounter road traffic. Sadly, far too many bike riders are involved in actual accidents and deaths. While we can argue that there should be more bike lanes, and drivers should be more bike-aware, bicycle users have to deal with the situation as it is today. If you can achieve your healthy cycle separated from traffic, it's a great form of exercise, but if you put your life at risk daily (not to mention spending all your time breathing in fumes from exhausts if you cycle in a city centre), it's arguable that the negatives outweigh the positives in health terms.

So, if the environment and the rides work for you, whether it's commuting to work or social riding with a club, a bike could be a great asset, but don't feel you are in any way an inferior exerciser if you opt instead for your own two feet.

LINKS:

D

• • • • • • • •

Duration

There is a lot of varied information on how much exercise is the minimum we need, but there is a reasonable identified minimum.

For many of us, with busy lives and a sedentary lifestyle, particularly those of us who have never enjoyed sport, the key question is: 'How little can I get away with?'

In the US in the first decade of the 21st century, a huge amount of effort was put into researching the required duration of exercise for a healthy life, but unfortunately the outcome was inconsistent across the many studies. However, somehow the experts condensed this contradictory information into a recommendation that we should undertake a minimum of the exercise gained from 150 minutes of walking or 75 minutes of jogging per week. Thankfully, this can be split into chunks rather than being done all at once.

There is some benefit to going beyond that bare minimum, but if you are feeling guilty at not being a sporting superstar, all the evidence is that by far the biggest benefits come from the difference between no exercise at all and that minimum level. Just that minimum brings your risk of dying prematurely down by about 20 per cent. After that, the incremental benefit is relatively small. Tripling the basic level only drops the premature death risk by another 4 per cent. So, there is some benefit from doing more – but you have to work a lot harder for it.

LINKS:
• **Frequency and timing** – page 130

F

· · · · · · · ·

Frequency and timing

**Does it matter whether you do all your
exercise in one go or break it up into chunks?
Is there an ideal time of day to exercise?**

All the evidence is that breaking your exercise into manageable chunks has no negative impact on the benefits, which is good when you consider that most of us probably aren't going to get through the recommended 150 minutes of walking or 75 minutes of jogging per week in one go (see **Duration**). The exercise period needs to be sufficient for the elevated heart rate to be sustained for some minutes, so it is unlikely you will want to break up the exercise into periods of less than fifteen or twenty minutes. Similarly, it is probably better not to go more than two to three days without exercise, putting a minimum number of sessions a week at two or three. Anything in between this and daily exercise seems acceptable.

One interesting observation is that the timing of your exercise can influence how effective it is at helping you to reduce levels of fat, if that is your goal. Specifically it's a good move to exercise before breakfast, as working out when you haven't eaten seems to use more fat and less carbohydrate. This approach seems to reduce the chances of gaining weight. (It has also been recommended that you then breakfast on eggs, as you are likely to consume fewer calories during the day after eggs than after a typical, carbohydrate-loaded toast or cereal type breakfast.) However, if you are only exercising for overall health, the timing is not an issue.

LINKS:
- **Duration** – page 129

K

· · · · · · · ·

Knuckle cracking

**Though most of us have probably been told
at some point in our lives that cracking
your knuckles will give you arthritis in old
age, there is no good evidence for this.**

It may not be what most of us would consider exercise, but some
people carry out a physical activity that seems to irritate others
around them out of all proportion to its offensiveness – cracking
their knuckles.

Almost inevitably the knuckle cracker will be told to stop, as
they are doing damage that will lead to arthritis in old age. But is
there any truth in this? One man, an American medical doctor
called Donald L. Unger, made it his life's work to find out. For over
60 years he cracked the knuckles of his left hand every day while
leaving the right alone. He has not suffered any difference in his left
hand as a result.

Although Dr Unger's dedication is quite remarkable, this isn't
of itself enough to persuade us, as one individual could easily buck
the trend. It's a bit like the way you will always find someone who
says: 'I've been smoking 40 cigarettes a day for 60 years and I'm in
the peak of health.' This could be true for one individual, but there
is plenty of evidence that smoking causes a wide range of medical
difficulties once a larger sample of smokers is examined.

Luckily, though, we don't have to just rely on Dr Unger's cracking piece of work, as more rigorous studies have confirmed the outcome of his one-man laboratory – there is no link between knuckle
cracking and likeliness of developing arthritis.

This is doubly confirmed by the mechanism of knuckle cracking, which is not as it appears a matter of abrasion of the joints, causing damaging wear. It seems more likely that the cracks come from bubbles in the fluid in the joints, which can be allowed to cause that impressive noise without any negative results.

However, whatever the cause, it doesn't prevent knuckle cracking from being surprisingly irritating for others. So maybe it's worth giving it a miss anyway.

O

· · · · · · · ·

Overload

**Basic, low level exercise like walking reduces
our mortality rate – but if you want to improve
physical fitness, you need to push yourself.**

Just moving regularly helps our chances of surviving, but it won't
improve fitness in terms of ability to perform a particular physical
task. To do this means pushing yourself – technically referred to as
'overload'. If you want this kind of benefit you should be exercising
sufficiently to get a little out of breath.

Overload doesn't mean pushing yourself to the absolute limit
for your whole period of exercise – that is likely to end badly – but
rather making sure that you are stretching yourself. So, for instance,
instead of just walking, you might walk for three to five minutes at
the fastest pace you can manage, then for a few minutes at normal
pace, alternating the two.

How much you need to overload is a moveable feast – the more
you do, the more you will need to do, as your abilities will improve
with practice. In effect, what was 'pushing yourself' becomes everyday, so you need to push further. Bear in mind, though, that this is
primarily about fitness in terms of improving your abilities, rather
than health. Even a comfortable level of exercise, provided there is
enough duration, will give you the health benefits.

There is increasing evidence that 'high intensity interval training' where bursts of extreme exercise of just a few minutes duration
are interspersed with low intensity exercise is better than traditional interval training or straightforward exercise routines. In one
extreme example, students riding a gym bike for 90–120 minutes a
day were compared with another group who went through a cycle

of 20–30 seconds at their absolute maximum, painful capability, alternated with a few minutes of rest. This was repeated four to six times. These three minutes of intense exercise gave them almost exactly the same increase in fitness as the 90 minutes to two hours of less intense riding.

A degree of overload also seems to have some health benefits. Studies have shown significant reductions in the risk of dying prematurely from some types of cancer in those who exercised fairly intensively (e.g. jogging) over and above those who undertook low impact exercise.

However, overload should always be brief – exercising to excess can result in both physical damage and also reduced ability to counter infections because the body is exhausted.

LINKS:
- Basic exercise and mortality – see **Duration**, page 129

R

· · · · · · · ·

Recovery

**It is received wisdom that we should perform
certain acts at the end of exercise to help recover,
from light exercise to ice baths. But do they work?**

There is no doubt that exercise has negative effects on our bodies. We ache after a good exercise session and can have inflammation from damaged tissue. However, a degree of soreness is not a bad thing, as muscles and tissues tend to rebuild stronger after being stressed – as long the situation is dealt with correctly.

If we look at the traditional responses to this soreness, most are useless or worse. It's often the case that after exercise we are encouraged to do a gentle exercise to 'cool down', the opposite of a warm-up – yet when studies have compared those who cooled down and those who simply stopped, there was no difference in the subsequent soreness.

Others apply painkillers like ibuprofen, nonsteroidal anti-inflammatory treatments which are known to be effective at dealing with muscle pain. Once again, this is based on myth rather than actual studies. Yes, these are painkillers, but taking them as a preventative doesn't work – in fact it tends to result in mild damage to the body, rather than advantages. The anti-inflammatories may also tend to block the rebuilding process after stressing.

Most of us have heard of the problems of lactic acid build-up in the muscles during exercise, and how a massage helps deal with this. Interestingly, there is some doubt whether lactic acid build-up is a bad thing, but even if it is, when the impact of massage was studied, it was found to cut down the blood flow to the muscles, and as a result actually to slow down the removal of lactic acid. As yet it hasn't been established if massages have any benefit in recovery after exercise.

And then there's the ice bath, famed as tennis player Andy Murray's approach of choice. It is supposed to reduce swelling and soreness, encouraging recovery. But all the evidence from actual studies, as opposed to anecdotal evidence, is that there is no speeding up of recovery and some evidence that the process actually increases the muscle pain felt later.

In effect, all these techniques are placebos, the sports equivalent of homeopathic medicine. They can make us feel that we are getting better, so we suffer less – and as such, may still be worth doing – but the effect is all in the head.

What has been shown to be beneficial in recovery is resting. The body does need a chance to recover while not being stressed. Ideally you should have one day a week when you don't do any significant exercise, particularly if the other days you are really pushing yourself. But having too long a rest period – several days – in intensive training does seem to increase the risk of injury.

LINKS:
- **Painkillers** – page 310
- **Warm-ups** – page 147

Resistance training

Also known as strength training, this form of exercise, as opposed to aerobic exercise, is valuable for strength and sports abilities but has fewer health benefits.

For many athletes, resistance training or strength training is an important part of the exercise routine. It's all about getting the muscles stronger to be able to deliver more power at the desired moments. It really isn't designed to increase overall health.

Nonetheless, there have been shown to be some health benefits from resistance training. Like aerobic exercise, weight training, for instance, has been shown to help control blood sugar levels and reduce the levels of fats in the bloodstream that contribute to heart disease. It is also valuable for those who are getting older and naturally tend to lose muscle tone, in the extreme making mobility difficult. Weight training can be valuable here.

However, for general health benefits, it is generally easier to make use of aerobic exercise like brisk walking, jogging, cycling or swimming, than it is to undergo resistance training. While it wouldn't do any harm to get some muscle work in to avoid that wasting as you get older (using a full bottle of water as a weight, or one of the strong elasticated strips used by physiotherapists is a great way to undertake some light resistance training), if you are only going to do one, the aerobic exercise is the one that is most likely to be sustained and deliver real health benefits.

LINKS:

- **Aerobic exercise** – page 121

Running

Perhaps surprisingly, the reason that running is a good form of exercise is because we aren't very good at it.

If we compare the sleek run of an athlete with the plodding walk of the person in the street, it seems obvious which is the more efficient form of motion. The athlete, it would seem, is conserving every bit of energy and pumping it into their movement, while the ordinary person is wasting energy all over the place. In fact, though, entirely the reverse is true. The fact is, unlike some animals, running is not our natural mode of locomotion.

When scientists compared the way we use energy when running with walking, they expected to find that some modes of running, probably when landing on the balls of the feet or the toes, would come out as the best. But they were surprised to discover that the walking mode is far more energy efficient – it takes less energy to cover the same distance walking than it does running, particularly when landing first on the heels. Unlike many mammals that use a spring action as they hit the ground to conserve energy when running, we are natural walkers. According to the study, we consume more energy when running than a typical mammal of our size would be expected to.

Now this is bad news for runners who want to use up the least fuel. But when running for exercise it is actually a positive – because at a run you will consume more calories over the same distance, and so will have a better balance of calories eaten to calories used up in exercise. 'Power walking' – pushing a walk well beyond the natural speed – is more like running in its impact. But for the ordinary walker, a run will increase the energy use rate.

LINKS:
- **Calorie intake** – page 17
- **Walking** – page 146

S

· · · · · · · ·

Sitting

If there's one thing we could do less of these days, it's sitting around. And there are genuine health benefits from cutting down the amount of time we spend doing it.

It might seem pretty obvious that sitting, doing nothing much, is one of the best ways to conserve calories – and it is. (It's also the best way to conserve water.) And that's fine if you are stuck in a desert with no food or drink. But most of us do far too much sitting. We sit at work, sit on the bus or in the car, and sit again in the evening, in front of the TV, computer or game console.

Take a moment to work out roughly how many hours a week you spend sitting – get a feel for the proportion of your waking hours. Unless you have an active job, it could be scarily high. But all the evidence is that too much sitting has real, negative effects on our long-term health and happiness.

There have been health scare stories that 'sitting is more dangerous than smoking'. This is an exaggeration (and also suggests that the two are exclusive, which of course they aren't). But there is reasonable evidence that sitting too much is bad for us in ways that go beyond the obvious fact that we aren't using up much energy, and so are more likely to consume more calories than we burn off. (Especially as it's easier to snack sitting than on the go.)

When we exercise, our bodies handle sugars differently. Our bodies convert carbohydrates to glucose to burn in exercise – when we are active, those glucose levels reduce more efficiently to healthy values than if we are sitting around. And if glucose levels remain too high, it can result in type 2 diabetes and heart disease.

The obvious solution (as usual) is to do more exercise, but a small scale trial, where workers who normally sat at a desk most of the day were asked to use converted desks where they could work standing up for three hours a day, has shown that blood glucose levels came down more quickly for those standing. (There was also a small increase in calorie consumption, though standing isn't hugely better than sitting in this respect.)

In practice, most of us can't necessarily work standing up and, though in principle we could watch TV in the evening standing, it's surprisingly hard to do (I've tried it). Standing works best with an active, rather than passive role. But we can still try to break up our sitting day.

Take a look through a typical day's activity. If you have large chunks of the day where you are sitting, look at ways to break it up on a regular basis. I have this problem as a writer, spending much of my time in front of a computer screen. So, I will try to make sure that, for instance, I walk the dog at two points in my working day, that I take a walk to the post office in the late morning … or even do a bit of hoovering mid-afternoon. If you can break up that sedentary behaviour with regular movement, you will be doing yourself a big favour.

Also, look for activities where standing or walking comes naturally. Many of us, for instance, have a tendency to stand or pace around when on the phone. Instead of suppressing the urge and sitting, give in and make use of it to get up and moving.

LINKS:
- **Calorie intake** – page 17

Stairs

**Almost wherever we go we are faced with a free,
easy-to-use gym that many of us stubbornly ignore.**

I am sometimes amazed by the number of people I know who are prepared to pay large sums every month for gym membership, but who ignore a free, easy-to-use gym that they come across all the time. As you might have gathered from the title, I'm talking about stairs.

Going up a flight of stairs (or preferably several of them) is a great form of aerobic exercise. It's gym equipment, handily laid on practically wherever you are, for free. And yet those gym members I know will usually catch the lift or take the escalator instead. Many of us encounter stairs in a number of places in our working day: in stations, in office buildings, in town centres. By simply making use of them, rather than taking the lift, you can instantly increase your exercise levels.

If you work on the twentieth floor, you might not want to walk all the way (at least to begin with) – so start getting off the lift two floors early and work your way up over time. Our bodies get used to exercise quite quickly, at which point just slip in another floor. You will probably find yourself arguing that you haven't got the time and need to take the lift. If that's really true, just arrive a few minutes early. But for most of us it is an excuse, not a genuine reason to avoid taking the stairs.

Of course you might not work, or, like me, you might usually work from home. Unless you live in a bungalow, you still have at least one flight of stairs available. Make more use of them during the day. Ideally have a reason for your trips up and down – do some tidying as you go – exercise is always easier to justify if it has a secondary purpose.

LINKS:
* **Aerobic exercise** – page 121

Stress and exercise

**Stress is a widespread problem and increases
the risk of heart problems and other illnesses.
A small amount of exercise can make a
big contribution to relieving stress.**

Exercise has a dual impact on stress, building physical fitness and
general ability to cope at the same time as bringing down levels of
chemicals in the body like adrenaline (epinephrine) that push up
stress levels.

One of the most common problems with attempting to exercise
to reduce stress is keeping it up. After a couple of enthusiastic weeks,
it's easy to start finding excuses not to bother. Make use of some
strong levers to encourage you to keep it up.

Firstly, make sure that you are doing it for a good reason. Ideally,
look for a way to emotionally blackmail yourself. Come up with
a reason like 'to stay alive for my children' or 'to make family life
better as I won't be stressed all the time'. Keep this reason in mind
when coming to the point where you are likely to try to weasel out
of the exercise.

Secondly, try to make the exercise something you enjoy. All
too many types of exercise are deadly dull, requiring some kind of
distraction like music or a TV to stop you from giving up. Look for
a form of exercise that will give you more than just working out.
For instance, if you can do it outside, take in the world around you.
Find other ways to add value to what you are doing by making it
social, or if you must plug in the earphones, go for an audio book
or learn a language, don't just work out to music.

Thirdly, if you are struggling to get your de-stress exercise in, try
to integrate it with stuff you would be doing anyway. Walk instead
of driving to the shops. Get off the bus a few stops early. Use the
stairs instead of the lift.

Don't underestimate the value of walking. Just getting off the couch and moving around provides a positive benefit for your health. And if you start to push that walking harder it can be surprisingly effective as exercise.

To make walking more effective for stress relief, consider an environment where there are no traffic-related stressors, like the countryside or a park. Don't consciously think about your problems. As a side bonus, you may well find that solutions to problems that are stressing you do pop into your mind as you walk – and if that's the case, don't ignore them, jot them down or make a voice note on your phone. Then carry on taking in the world around you.

If you do take your phone, turn it to 'do not disturb' mode. Otherwise you are carrying a stressor with you all the time.

LINKS:

• **Walking** – page 146

Stretching

If there is anything that divides people who take exercise seriously and the ordinary person it's all those complicated stretches, yet there is good evidence that they don't work.

Since time immemorial it has been exercise gospel that it was important to do stretches during a warm-up before serious exercise to avoid damage. The routine will normally involve 'static stretching' – stretching a muscle to its extreme and holding it for a number of seconds before relaxing it. This process may go on for as long as ten or fifteen minutes, after which the would-be athlete invariably feels more supple and ready for their exercise. But there's a catch.

This whole stretching routine was based on what felt right, rather than on science. When a group doing stretches were compared with a group who simply rested before exercise, the stretchers performed worse. They covered less distance in a run and consumed more calories and oxygen. Their ability to run had become less efficient – which, when you think about what is being done to the muscles in stretching them, is not entirely surprising. Taut muscles are more likely to provide an energy-reducing elastic response than relaxed muscles. Yet stretching is so much part of the exercise culture that no one had thought to question it.

If you still do warm-ups involving touching your toes, stretching out your legs to the extreme or similar, you are not doing yourself any favours (unless the sport you are involved in specifically requires flexibility, like gymnastics). The decrease in muscle strength from this static stretching can be as much as 30 per cent.

You may think that your stretching is reducing your risk of injury, even if it doesn't improve performance – but this is a myth too. In large studies of those undertaking exercise, there was no difference in injury between those who stretched before and those who didn't. Equally there has been no benefit demonstrated from static stretching as part of a recovery exercise.

It is still worth warming up – but stretching is not the way to do it.

LINKS:
- **Recovery** – page 135
- **Warm-ups** – page 147

Swimming

**A great, all-round way to get exercise,
if it's convenient to do so, swimming is
well worth adding to your routine.**

Leaving aside the obvious health benefits of learning to swim if only to survive if you fall in water, there is no doubt that swimming is a very effective means of exercise that has much to recommend it, getting a wide range of muscles going and providing resistance to practically any movement in water-based exercises to help recovery after an injury. In exercise health impact terms, swimming is similar to a brisk walk, but gives more of the muscles a workout.

The only real downside to swimming is that it takes up a considerable amount of time in reaching the swimming pool (unless you have one of your own), getting changed and so forth – compared with taking a walk out of your front door, it is high maintenance – and swimming is not usually a free activity.

There are small health risks, like gaining a verruca, but there is no evidence of real problems with chlorinated water that some worry about – certainly not for any normal exposure levels – and it's much better than swimming in a vat of bacteria and algae, as you are likely to be doing without the chlorination.

LINKS:
• **Walking** – page 146

W

· · · · · · · ·

Walking

**Walking is often underestimated as a
form of exercise, and yet, done properly,
it can give real benefits and can be easily
integrated with your daily routine.**

The trouble with walking is that it sounds too leisurely. Yet even a
stroll has positive benefits over sitting around, and if you push your
walk to the fastest pace you can manage it is also a very effective
aerobic exercise. What's more, there is less risk of injury than there is
from jogging and running, and there are opportunities to combine
a walk with something practical like exercising the dog or a trip to
the shops, enjoying the scenery or collecting firewood.

In practical terms, fast walking uses more energy per minute
(for any duration) than riding a bike (unless you are really pushing
the bike to the extreme). To take your walking from a stroll to serious exercise, begin with a few minutes at your normal pace, and
then start to push it up. Most people speed up by lengthening the
stride – as well as doing this, increase the frequency of your paces.
Push it as far as you can, then throttle back a little. You should be
walking fast enough that you really feel the muscle use and become
a little out of breath. Most people can achieve around four miles an
hour – and with practice you will gradually increase your speed
and the exercise value.

Most of us have opportunities to walk that we ignore. For
instance, if your workplace is less than four miles from home, why
not walk to work once or twice a week? The immediate reaction is
that this would take up too much time – but it is a maximum of an
hour each way for a single round trip – and you might spend up to

half of that sitting in a car in a rush hour jam. Many people happily spend two hours a week or more in the gym. But walking has many more benefits. Not only do you get the exercise, you save on travel costs, benefit the environment by not producing exhaust emissions (and save money if you have a gym membership to cancel). What's more, whether it's an urban or rural environment, there's much more to see and experience on a walk than either in the car or on a running machine in the gym.

LINKS:
- **Cycling** – page 127
- **Duration** – page 129
- **Running** – page 138
- **Stress and exercise** – page 142

Warm-ups

Unlike stretching, there is good evidence that other warm-ups are beneficial before taking exercise.

It might seem that the debunking of the need for stretching means we can get rid of warm-ups before exercise altogether. But there is a lot of evidence for the benefits of the right kinds of warm-up. Warming up has been shown to improve performance in pretty well any sporting activity. There is a proviso that a lot of this evidence is not very good, in a scientific sense. But the consensus at the moment is that the best approach is to warm up first.

The main benefit seems literally to be warming up the body into a more active state, heating the tissue, which seems to give better performance and lower risk of injury. The warm-up does not have to be fancy. So, for instance, before running, a brisk walk would provide an ideal warm-up – and before a brisk walk, walking at a comfortable pace. No contortions required.

For a more significant warm-up, the exact requirement will vary from sport to sport, but the main things are to avoid static stretching and undertake warm-up exercises that get the same bits of the body moving as will be doing the work in the sport. The likes of star jumps, skipping and bottom kicking, for example, are all beneficial where relevant to the exercise coming after – it's only static stretching that is worse than useless.

There are ways to get warm-ups wrong. If you warm up, then sit around before the actual exercise, it can be worse than not warming up at all, resulting in more stiffness from the period of inactivity than would have been the case with no warm-up. Similarly, a warm-up that is too close to the actual activity in the level of exercise can be counter-productive. The warm-up has to be significantly gentler or you can simply tire yourself out before getting to the actual exercise.

LINKS:

- **Stretching** – page 143

BRAIN

. . . .

The human brain is an amazing organ – the most complex object known to science. In around 1.5 kilograms of matter it houses 100 billion special cells called neurons. The complexity of the brain is in the way those neurons can be connected together, which is how this remarkable organ stores and manipulates information. At any one time there are likely to be around 1,000 trillion connections in your brain, and the number of possible connections far exceeds the number of atoms in the universe.

Scientists have put a lot of effort into getting a better understanding of how the brain works. There are still major aspects of our brains that aren't at all understood – we don't know, for instance, what consciousness really is or how it comes about – but it has been possible to map various mental activities to parts of the brain and to begin to understand the complex electrical and chemical processes that enable it to function. With this in place, scientists are better able to give guidance on possible ways to train and exercise the brain, some of which offer real possibilities for improvement, while others have proved totally worthless.

Before starting on an instruction manual for using the brain better, it is useful to know that we really each have two brains. Although the brain looks like a single object, it is actually divided into two totally separate halves, joined together only by a thick bundle of nerves at the back.

The two sides are mostly responsible for the opposite sides of the body – so the left side of your brain mostly looks after the right hand side of the body and vice versa – but only 'mostly'. So, for instance, though your left eye is mostly connected to the right-hand side of the brain, some of the signals from it go to the left side, to help with processing and working out three-dimensional spacing.

It was once thought that there was a very clear distinction between the two sides. The left side was thought of as the logical side, the business brain. This was the side that dealt with words and worked through things in a careful, sequential manner. The right side, by contrast, was arty, handling the overview, spatial awareness, colours, music and images.

Again, there is an element of truth to this – the brain certainly has two distinct modes of operation that correspond to this 'left'- and 'right'-brain thinking – but when scans have been done using an fMRI scanner, which highlights exactly which parts of the brain are active during different activities, it was found that there was always some overlap. Left- and right-brain thinking aren't wholly limited to the relevant side of the brain that gives them their label.

Whether thinking of yourself, your children or elderly people you know, it's a no-brainer to discover how to make the best use of the little grey cells.

A

· · · · · · · ·

Ageing

Most of us get the feeling that our brains become less effective with age, but is this really true?

Leaving aside degenerative illnesses that can reduce our mental capabilities as we get older, it has always been felt that our brains are no longer functioning as effectively by the time we reach retirement age. It used to be said that brain cells start dying once we get to adulthood, meaning that we lose more and more as we get older. This has been shown to be an over-simplified picture. Yes, we do lose brain cells – but unlike early assumptions, some new ones are also created. Plus, the brain is such a flexible organ that it isn't clear that simply losing brain cells is enough to change our capabilities.

A study published in 2012 gave a clearer picture on ageing, though there are some provisos. This showed that there was some deterioration in mental capabilities from the age of 45 and significantly more by the time participants were 60. (It didn't tell us anything about younger people, as none of the participants, who were monitored over a ten-year period, were younger than 45.)

Unlike many studies on the brain, this was a large-scale survey, so it does seem to be reasonably robust in its portrayal of a 3.6 per cent decline in those initially aged 45–49 over the ten years, and a 7.4 per cent decline in those initially aged 65–70. But we do have to be a little careful about drawing too many conclusions. Specifically, the 7,000 plus participants were all British civil servants, which is not a typical cross-section of the population.

What was being tested was 'cognitive reasoning', in terms of maths, verbal fluency and vocabulary. Although the study does show a decline, it is interesting that it isn't by a huge percentage.

Quite possibly it is one that would not be noticed subjectively by those taking part – and just because individuals decline a little in the face of such a test does not mean that they have ceased to be able to take part in their chosen activities well. In fact, it would be perfectly possible to have this kind of decline in test ability and an improvement in real-world abilities. In the real world, for instance, knowledge and learned abilities can both increase our capabilities as we get older, but these were not tested.

It is also worth noting that there are certainly activities that help to slow down decline in some of these abilities, which will be covered later in this section (see links). It could be that individuals who undertook such activities would experience less decline – or it could be that the civil servants in the study were more likely than most to have regular 'brain stretching' activities and as such would experience less decline than a typical person. This study is a useful starting point, but more work is necessary if we are really to get a feel for how our mental abilities change over time.

LINKS:
- **Brain-boosting habits** – page 156
- **Brain training for fading capabilities** – page 162

B

. . . .

Baby brains

**We know that the human brain goes
through a huge amount of development
in the first few years of life – but what
can we do to ensure the best outcome?**

When a baby is born, its brain is already surprisingly well formed. The majority of the neurons – the billions of cells that provide the main foundation of the brain – are already in place. But at this stage there are very few connections between neurons, and it is these connections that enable us to think and to remember. Connections between neurons are made and broken throughout our life, but 80 per cent of the basic pattern of connectivity will be set up in the first two years of life.

There is good evidence (primarily from animals) that the effectiveness of the building of this initial framework is strongly influenced by the experiences the baby has in these early years.

Some of the earliest links to be made are those involved in the interaction between the senses and body movements. Young babies spend a lot of time not just making movements, but also watching those movements, coordinating vision and the ability to control muscles. There has always been a feeling that giving babies rich sensory experiences they can interact with, from textured books to play environments and plenty of personal interaction, helps with this development, and this has been borne out by research.

A little later come activities that help the baby brain cope with the physics of the world. How things move and fall. How something disappears when it goes inside something else. What happens when you touch something or push it. You may get thoroughly fed up with your baby throwing things on to the floor, but one of the reasons they find this such an enjoyable exercise is that they are performing baby science experiments, learning how things move when they let go of them, building an internal 'model' of the world that will help the older brain cope with many kinds of environment.

These developments of the baby brain continue with a very important stage where they learn to link what they sense with personal experience – so, for instance, realising that seeing someone else drop something is essentially the same thing as dropping it themselves. This expands the model of the world that is forming in the baby's brain to include other influences than his or her own.

A huge amount of early development, both in physical and social skills, comes from imitation. The baby's carers naturally tend to encourage this with repeated sounds and gestures – all the time the new connections are being forged between those neurons to help the baby function independently. The brain is learning how to interact with the world by observing others. This is why social contact, as well as physical stimuli, is so important in this formative period – and why babies brought up in circumstances where they have very little social stimulation can be seriously disadvantaged.

An obvious example of the way experience builds patterns in the brain is in language development. We are born with an inbuilt 'language ability' but no specifics. It is through social interaction, initially through repetition of noise patterns and later through lots of opportunities to put the sounds of language together with other sensory stimuli, that the baby begins to communicate. It's hardly rocket science, but it really does help to talk to a baby as much as possible.

It is here where one of the few examples of the often-heard idea that TV is 'bad for you' has evidence to back it up. Babies can obviously pick up stimuli from TV; so letting them watch it isn't totally negative, but TV lags far behind real-world stimuli because there is no feedback. When the baby drops something on the floor, it gets a response. When we talk to it and it responds, we react. In the real world there is always feedback to reinforce and build those essential brain connections correctly.

TV offers no feedback. So the recommendation is, as much as possible, to avoid unsupported TV viewing in the first couple of years, if only because time spent concentrating on the images on the screen is time that could otherwise have been occupied in building that mental model. There's nothing wrong with babies experiencing TV, but it should always be accompanied by human interaction to provide the feedback.

By comparison, looking at a picture book with a baby offers much more opportunity for interaction and feedback. We naturally tend to reinforce words with repetition, linking words to images, just as we do in the real world when, for instance, pointing out a passing dog in the park – soon the baby will be pointing to a dog and telling us about it. The book provides an interesting development of the mental model of the word, because it helps the baby comprehend the idea of representation. A picture of a dog in the book isn't a dog – yet it represents it.

This is a great stepping-stone to the idea that the word 'dog' in a book also represents the furry reality. Looking at a book with a baby and increasingly reading to it is one of the best developmental acts for the brain in terms of having good research to back up the benefits.

Brain-boosting habits

Attempts to improve the brain's capabilities often rely on a significant amount of effort, but it has been suggested that two everyday factors can make a difference.

An analysis of the performance of the 40 million members of the online brain-training site Lumosity provides some interesting food for thought on simple boosts for brain effectiveness. Subject to the scientist's favourite warning 'correlation is not causality', studies of the database reveal that participants who had seven hours of sleep a night and had one or two alcoholic drinks a night did better than those who moved away from these figures in either direction.

That's quite a claim. Not only, as common sense might suggest, did the seven-hour-a-night, one or two drinks people do better than those who slept less or drank more, they also did better than the eight-hour snoozing teetotallers. We'll come back to why this may be the case in a moment, but the most dramatic difference seemed to come from physical exercise. Just one burst of good exercise a week was enough to push participants' speed of performing Lumosity's tests up by nearly 10 per cent and saw enhancements in mathematical problem solving and (to a lesser extent) spatial awareness.

As you'll see elsewhere (**Exercise and the brain**), there is good evidence that physical exercise contributes to mental acuity; so that part is no surprise. But why would exceeding seven hours of sleep or not drinking alcohol at all have a negative result? If we take it at face value and look for a cause, then it is possible that too much sleep could have a physiological impact that leads to lower brain effectiveness. A study has shown that different sleep lengths can result in genes being switched on or off, and it is entirely possible that, as well as this influencing susceptibility to various disorders, it could also impact on the brain's activity.

Similarly, we could argue that while even a small amount of alcohol reduces our ability to react quickly and, for instance, drive a car, it can also provide the kind of relaxed ease of mental connections that is helpful with the kind of tasks that could come up in some of the Lumosity tests. It isn't a universal aid to brainwork, but a small amount of alcohol may aid creative thinking.

However, it could equally be the case that these advantages are not caused by the sleep duration or drinking, but rather it could be that the kind of person who has around seven hours of sleep a night and one or two drinks is more likely to be, say, a successful, experienced but not too old person, rather than, say, the typical long-sleeping student or the person who is teetotal because of medical problems that make it harder for them to operate a computer keyboard. I am not saying those *are* the actual causes; I am just giving examples of a secondary cause that might link the two. If this were true, sticking to seven hours of sleep or knocking back one or two alcoholic drinks a night would not cause any improvement in brain function.

LINKS:

- **Brain training for fading capabilities** – page 162
- **Brain training for intelligence** – page 164
- **Exercise and the brain** – page 169
- **Music** – page 181

Brain food

For years we were told to eat fish because it was 'brain food' – but can consuming anything improve your mental capabilities?

It's hard to avoid product advertising suggesting that this or that supplement will make your brain (and especially the brains of your children) a super-effective vehicle for superior thinking skills. There's nothing new about this. Fish, for instance, has long been labelled as 'brain food'. But of late, everything from those ever-present superfruits to omega 3 oil supplements has been flagged up as the must-have for a healthy and enhanced brain. The other day I even saw in the supermarket hens' eggs labelled with the claim that they would make you brainier, just because the hens had been fed on a particular diet.

One of the problems here is that there have been occasional trials that seem to show a benefit in, say, feeding cod liver oil to school children. But like so many of the trials over-hyped in the press that lead to the confusion that *Science for Life* is here to disentangle, these have been one-off studies, often with small numbers of participants or badly handled. In short, they are the classic kind of trial that can't be taken as worthwhile evidence.

This might be a bit of a surprise because one trial involving 5,000 children sounds very impressive; it supported the benefits of fish oil and is still widely referenced in the press. That 'trial' was a fiasco. Durham Local Education Authority approached a fish oil supplement company, asking if they could run the experiment. The trouble is that there was no control planned – no attempt to compare children with and without the supplement. All they intended to do was see how the children did against expected outcome.

As set out, this Durham experiment would have been totally valueless. To make matters worse, the psychologist and head of education involved were featured on TV before the trial, saying they *expected* positive results. As soon as someone tells you what they expect the result of this kind of trial to be ahead of undertaking it, it's not science. Scientists know that if you go into a subjective trial expecting a particular result it's almost impossible not to bias the outcome in your preferred direction.

Three thousand children started the experiment, but 2,168 dropped out. What results were discussed publicly involved *some* of those who stayed in – but this is 'cherry picking' – a well-established flaw in trials that makes them useless. As soon as you start picking out the results that are favourable to your hypothesis, you can make a trial say anything.

Other trials have involved proper double blind controls with placebos – but on a much smaller sample of children, and on children with issues like ADHD, so they don't give any indication of outcomes for children in general. Even here, reporting suggests much better outcomes than actually occurred. At the time of writing, there is still a BBC website reporting on the successful outcome of a trial, telling us about Elliot, aged nine: 'Over the past year, a dramatic change has taken place in Elliot. He has soared through the Harry Potter books and now heads to the library after the school bell has sounded.' Unfortunately, when that report was written the trial results weren't available, and they proved mixed at best.

More recently, though, there have been sufficient large, good quality trials that it is possible to take an overview of whether we can enhance our brainpower by changing our diets – and with a couple of specific exceptions, the answer is 'no'.

The biggest disappointment is the omega 3 oils responsible for those school trials and the supermarket's brain-enhancing eggs. Not only do the respected Cochrane reviews fail to pick up any significant benefit, there are a couple of negatives. A review of trials in older people shows that the oils can cause mild gastrointestinal problems for some users. Another, more worrying outcome is that trials of use of fish oil supplements by pregnant women suggest a possible negative effect on future cognitive ability of their children – though this result is also not supported by enough trials to be considered definitive.

There seem to be just two things we consume that have a strong indicator of a positive effect on brain capability, one well reported and the other that may come as something of a shock. The no-brainer is feeding babies breast milk. As well as the recognised physical advantages (see **Breast milk**), there is a small but significant increase in mental capability. This is less than 5 per cent, so people (like me) who were bottle-fed need not panic – but the advantage is there.

The other positive thing to consume is a beverage that many of us tend to think of as bad for us … coffee. This is not just about getting a caffeine boost to make us alert – comparisons of coffee (which contains many complex chemicals) with caffeine alone make this clear. Studies have shown that coffee both improves mental capacity in the middle aged and reduces the chances of developing dementia in older age. As always with stuff we consume, there's a mixed message. Coffee is the most carcinogenic component after alcohol in a typical diet (though it is still very low risk). But it emphasises that the substances with a reputation for enhancing brainpower aren't necessarily the effective ones.

LINKS:
- **Brain training for fading capabilities** – page 162
- **Brain training for intelligence** – page 164
- **Breast milk** – page 266

Brain Gym

Brain training provides mixed results, but many UK schools have used a commercial system known as Brain Gym. This combines some sensible basics with other aspects that lack a scientific basis.

If your children went to primary school in the 21st century in the UK they may have experienced Brain Gym, a commercial methodology for improving thinking.

Some basic aspects in the Brain Gym process are simple, well-established factors – like it being good to take regular breaks when learning, the benefits of exercise or the value of basic breathing exercises. However, much of the rest of the process has been shown by Dr Ben Goldacre to have no basis in science, despite making use of pseudo-scientific terms in its literature.

Specifically, Brain Gym has a strange attitude to water and odd rituals. The claim is made that it's important to drink a glass of water before the activities, and to hold the water in the mouth to transport oxygen to the brain. Drinking fluids is certainly a good thing – but the Brain Gym approach to this doesn't make any sense. To begin with, it specifies that only pure water satisfies the body's water needs, where in fact we take in water from almost all food and drinks from squash to tea. And holding water in the mouth won't transport oxygen to the brain. What's more, as is described in the **Hydration** section, we are rarely dehydrated in developed countries.

There are also special massage-like movements in the Brain Gym routine, which are said, for instance, to stimulate the blood flow in arteries to carry more oxygen to the brain. Again, these make no physiological sense.

At the time of writing, schools are making less use of Brain Gym than they were before the recession due to funding limitations, but many teachers and pupils have been persuaded that there are benefits in a regime that makes no scientific sense.

LINKS:
- **Brain training for fading capabilities** – page 162
- **Brain training for intelligence** – page 164
- **Exercise and the brain** – page 169
- **Hydration** – page 63

Brain training for fading capabilities

**We all lose mental capacity to a degree
as we get older, but there is a simple way
to hold on to as much as possible.**

There are plenty of products that claim to exercise the mental muscles, keeping the brain fit and effective. Although usually advertised by forty-somethings, these are targeted at older users who feel that they are losing facilities like memory, and want to keep as much as they can of their abilities. We all suffer a degree of memory loss as we get older, so it is helpful to know if there is any benefit to be had.

There is some truth in the adage 'use it or lose it'. If you really do nothing with your brain, sitting and vegetating, there is good evidence that you will be less able to function mentally. So, it is important to socialise and undertake activities that keep the brain active (socialising is enough, in fact, on its own) – but all this does is stop things going downhill. There is no evidence that it 'trains' the brain in any sense to be more effective. You will still suffer memory problems, for instance, but it will keep the decline to the minimum.

According to researchers, the most important tool by far in keeping your memory effective is overcoming self-doubt. If people give in to apparent loss of memory, it is then that they start down the slippery slope. They don't use modern technology, because it is beyond them. Perhaps they give up reading books or doing crosswords. They hand their bills over to others to handle and don't bother to socialise. Every rejection of ways that will keep the memory active, every abrogation of personal responsibility, will help in memory's destruction. The benefit of 'using it' comes from any mental activity, though, rather than specific 'brain training' exercises.

Is there anything to be gained from using 'brain training' software or taking supplements to enhance your ability to retain memories? In reality, memory isn't like a muscle that can be built up by regular exercise. There are techniques that can be used to undertake memory tasks like remembering names or a shopping list. And they work for that kind of activity. But they are no help in the sort of memory we are desperate not to lose – the recall of what has happened in our lives and to our loved ones.

Occasionally you will also see the old chestnut that 'we only use 10 per cent of our brain', implying we need training to access the rest. This has no scientific basis whatsoever. The suggestion of 10 per cent use dates back to a time when we couldn't see the brain at work; now we have fMRI scans it is very obvious that this simply isn't true. There never has been any basis for the claim – and in scans, the only bits of the brain that are inactive are dead bits.

As far as supplements to keep brains functioning well go, they work on reverse logic. It is true that a shortage of some nutrients such as vitamins E and B1 can result in memory problems. But most of us are not short of these vitamins – and giving us more than we need does not make the memory any better. A shortage is bad, but that doesn't make excess good.

So, by all means enjoy 'brain-training' games – they can be fun – but don't expect them to make a lot of difference to your memory that couldn't be done by having a chat with a friend or reading a book.

LINKS:

- Brain training that really works – see **Brain training for intelligence** – page 164
- List techniques – see **Remembering lists** – page 192
- Name techniques – see **Remembering names** – page 195
- Number techniques – see **Remembering numbers** – page 197

Brain training for intelligence

It is possible to give a slight boost to intelligence with some kinds of mental activity – but it's hard work.

There are specific mental exercises that do seem to have a positive influence on one aspect of the brain's working. The training exercise, usually called 'double N-back' involves making decisions based on a piece of information several stages back in a chain of numbers, pictures or words. So, for instance, you might see the words 'five, six, eight, four, nine' flashed up in sequence on a screen, then after they have disappeared you might be asked to take three away from the last number but one. The 'double' part means that the subject has to work on two different chains of information at the same time. The exercise is painfully difficult, requiring intense concentration. At first it seems almost impossible, but over time, pretty well everyone improves.

Studies do show this type of exercise produces an enhancement in working memory, an essential component in what we generally think of as intelligence – certainly important in problem solving. It ought to be stressed that the improvement is best described as 'small but reliable'. It isn't going to turn the average person into a genius. But it does seem to have an impact across the board in everything from conventional academic studies to more sophisticated problem solving. And there is good evidence for more of an impact if the participant is particularly in need of assistance in intelligence-related tasks.

By comparison, a study in the UK pitted over 11,000 people against a battery of intelligence tests before and after a six-week programme of using brain-training exercises similar to those used in popular apps like *Dr Kawashima's Brain Training* on Nintendo. It ought to be stressed that the makers do not claim that these apps increase intelligence, but consider them purely as entertainment, though with titles like that it is easy to become confused. The result was very interesting. Practice at the puzzles and challenges makes participants better at doing the puzzles, but does not lead to enhancement in any other tasks, even if they appear to depend on the same type of cognitive activity.

If you would like to have a try at N-back, there a number of free programs online, including at www.soakyourhead.com. You can also find slicker, paid-for versions at www.lumosity.com.

LINKS:

- **Music** – page 181
- Use it or lose it – see **Brain training for fading capabilities** – page 162
- Working memory – see **Short-term memory** – page 200

D

. . . .

Drugs for mental performance

**Drugs that enhance mental performance
(with mixed results) have been available for a
while, but the surprise is that a villain of the
drugs scene may be a positive contributor.**

Brain-enhancing drugs cause much controversy. Even if we could produce a safe drug that would have a marked beneficial effect on mental ability, would it be politically acceptable? Would we see doping tests for TV quiz shows and random drugs testing in school exams? As it happens, this concern is not one for the present, as while there are drugs that seem to have some benefits, they also produce considerable side effects that are often totally out of proportion to any benefits.

Oddly enough, the one drug that has genuine, measured benefits and is already widely used is one that governments spend vast sums of money attempting to persuade us to avoid. That's nicotine.

There is good evidence that nicotine both increases cognitive ability and can help hold off neurodegenerative diseases like Parkinson's. It is widely available over the counter. And yet we do everything we can to prevent its use. Of course this is understandable. Tobacco kills many thousands a year. The last thing we want is a single person taking up smoking to enhance their thinking ability. But it isn't nicotine itself that causes the problems for smokers, it is the tar and poisonous chemicals that are also inhaled with cigarette smoke.

Surely, though, nicotine is hardly a good thing? At the time of writing, there is considerable controversy about e-cigarettes, which deliver nicotine without the other nasty components of the real thing. As I heard an expert say on the radio the other day, the problem with nicotine is that it is highly addictive. Which is strange, because there is no good evidence that it is.

It seems from countless trials that tobacco's undoubtedly fierce addictive properties are due to the combination of nicotine and a range of other powerful organic chemicals. On its own, these trials suggest, nicotine has no significant addictive effect. But it does lock on to receptors in the brain, enabling the substance to make us more alert, calm and less susceptible to dopamine-related issues like those afflicting Parkinson's sufferers.

Does this mean we ought to ease up on e-cigarettes for non-smokers? No. There is evidence that they can produce a greater chance of taking up actual cigarettes, particularly among the young. But it is arguable that alternative nicotine products could provide actual benefits in both mental acuity and reducing the impact of neurodegenerative diseases, and should be given more consideration by the medical community.

LINKS:
- **Brain food** – page 158

E

· · · · · · · ·

Environment for thinking

**The location where we try to think and
have ideas can have a significant effect
on our ability to think effectively.**

There is clear evidence that some situations and environments are
better for thinking than others, and yet all too often we don't do
anything to make things better.

One problem, particularly when dealing with the need to think
in a work or organisational context, is the general enthusiasm for
group working. (If you doubt this enthusiasm, compare the implications of an annual report that says 'she is a team player' with 'she is
a bit of a loner'.) Groups are great at taking an idea and improving
it – but groups don't come up with ideas in the first place, individuals do. So, if you are attempting to solve a problem or come up with
new ideas in a group, it's a good idea to split up individually first,
have ten minutes thinking alone, then come together.

Something that is pretty well universally observed is that a
good way to think is to go for a walk, ideally without a mobile
phone or other distractions. This is widely known by writers and
other creative thinkers, but has recently been supported by the
work of psychologists at Stanford University, who have demonstrated that during walking and for a short time afterwards there
is a boost in creative thinking ability. This almost always reaps
rewards in terms of ideas when compared with sitting at a desk
or at the kitchen table.

The main components of the 'thinking environment' seem to be
to get away from distractions and direct communication with other
people, to involve some movement and not to think too heavily

about finding a solution, but rather to let ideas flow, coming and going in the brain. It is also essential to have some mechanism to record them, as those ideas can be ephemeral, and as soon as you try to hang on to one idea, you won't be coming up with any new ones.

LINKS:
- **Exercise and the brain** – page 169

Exercise and the brain

Ever since ancient times we've heard the saying '*mens sana in corpore sano*' (a healthy mind in a healthy body) – and research confirms this, to a point.

There is nothing new to the idea that physical exercise is good for mental capacity (though the intellectual contributions of most top league footballers remind us that we are only talking about taking some exercise, not having the physique of a professional athlete). There can be little doubt that regular exercise has both health benefits and helps the brain. But what is surprising, as research has become more detailed and repeated to increase certainty, is that not all exercise is equal as far as mental abilities are concerned.

When comparisons were made, for instance, between mild aerobic exercise (for instance, fast-paced walking) and a regime of stretching and muscle toning, the aerobic exercise was found to be the one that impacted the most significant aspect of mental deterioration as we all get older – the ability to switch attention quickly and juggle multiple tasks. In a way this makes sense, as the stretching and toning primarily benefits the musculature, while the aerobic exercise has an impact on the cardiovascular system, with a more direct influence on the brain. There is also some good evidence of benefits for memory capability in children from aerobic exercise, suggesting those much-hated PE sessions are a good thing after all.

However, this is one of those areas where there are mixed results, and there are also less detailed (as yet) studies showing that resistance training with weights can provide benefit for the elderly with basic mental functions, particularly for those who are too elderly or infirm to cope with an aerobic exercise even as easy as fast walking.

LINKS:
- **Aerobic exercise** – page 121

F

. . . .

Finding things

**Why do we spend ages looking
for something and then find it ...
where we've already looked?**

If your house is anything like mine, you will spend a fair amount of time hunting for things, usually something you have put down, then forgotten where you put it. The bizarre thing is that you will often find it somewhere you have already looked. However, this isn't as strange as it seems.

Your brain is always taking short cuts to deal with the vast amount of information that comes in through your senses. It doesn't see the picture of the world you 'see' – instead it has modules for matching shapes, finding edges, identifying colour and so on. The movie-like picture of the world you think you see is just a construct, made up by your brain. By using a sort of pattern matching short-cut, your brain can manage to hunt for something as your eyes flick around the room. But the price for this quick action is that it won't actually be looking for the specific object in all its glory but rather 'a red rectangle' or a 'shiny circular thing'.

I ought to stress, by the way, that this is not literal, but the search process works at that kind of level. If, for instance, you are looking for a gold ring and it is at the wrong angle or partly obscured so it doesn't appear to be circular, that can be enough to prevent the pattern matching from triggering. Your eyes are pointed at it, but the modules in your brain that handle images haven't picked up on the specific trigger of a 'shiny circular thing' you were looking for. Then you take another look, perhaps from a slightly different angle, and suddenly everything clicks into place.

I

. . . .

Idea prompting

There's a reluctance to think that science can help us with such a human capability as creativity – but there is good evidence that there are techniques we can use to help us have ideas.

Before I became a full-time science writer I spent a lot of time helping businesses with creativity, something they all want but pretty universally struggle with. There are broadly three aspects of coming up with ideas that have good evidence to back them up.

The first is to be aware that we are all creative. Some industries (advertising, for instance) think there are 'creatives' and the rest. But it's not like that. Yes, some of us are better at coming up with ideas and better at understanding the need for an idea, but everyone is creative. Everyone can come up with a good idea. That's essential to know because most of us doubt our own creativity, so will tend to suppress our thoughts, particularly when talking with others, to avoid looking stupid. If you think you aren't creative, you won't mention your ideas.

The second consideration to help have ideas is that we aren't good at original thinking under pressure. The brain is a self-patterning system, which means that well-trodden pathways are the easiest to use. When you are under pressure, you tend to fall back on thinking the way you've always thought, rather than coming up with anything original. So, to help new ideas develop, it's a good idea to get away from pressure – and that includes thinking too intensely about the problem. Going for a walk or a drive, or relaxing in the bath – anything where your mind can wander and make new connections – is a far better way to come up with ideas than sitting at a computer or at the kitchen table with a piece of paper, trying to come up with original thinking under pressure.

Finally, since the 1950s a lot of work has been done both by psychologists and in advertising companies on mechanisms to support the production of new ideas. There are many ways to do this, but they all come down to the same fundamental approach. It is difficult to have new ideas when you are bound up by the constraints of your experience, of how things are normally done. So, what you need is a push to get you to look at your problem (or need for an idea) differently – to start from a different place. This is what these 'creativity techniques' do. Psychology studies have shown a marked benefit in the ability to generate new ideas when given some form of random stimulus in this fashion, often using a randomly selected word.

Let's consider a very effective technique quickly. The technique works by picking a picture at random. It needs to be unrelated to your need for an idea, and it should have plenty of detail in it. Although words are often used in academic trials, because they are easier to control, practical experience suggests that a picture provides a richer source of idea stimulation. Jot down what you are trying to achieve; then put that to one side. Look at the picture and note down what it makes you think of – anything it reminds you of, anything it suggests. Do this as a series of quick one-liners until you have around ten.

Now put the picture aside and come back to your need for an idea. With that in front of you, look at each of the associations you noted for the picture. What does the association suggest that could be a new idea that would help with your requirement? Some associations might generate a handful of ideas, some none at all. When you have a list of ideas, see if there are ways to combine them or improve them so they are even more effective, before choosing one or more to use.

I've used this 'random picture' technique thousands of times, and it is reliably effective at generating a new way of looking at a problem or a need for an idea.

L

. . . .

Left/right brain

The brain is physically two separate halves only joined at one point. Although the traditional left/right-brain thinking picture is too simplistic, it still has a relevance to the way we try to have ideas and think creatively.

Although we speak of the brain as a single organ, it is, to all intents and purposes, two separate halves (or hemispheres), joined at the base by a bundle of nerves called the corpus callosum. The two halves of the brain have different personalities. When the two sides are disconnected, either accidentally or in an operation used to treat certain conditions, they can have, for instance, totally different ideas on what the person wants to do as a career, on religious beliefs or on political inclinations. It is possible to discover these differences because the most important components of speech handling are dealt with by the left side, and of writing by the right side. So, a person with split hemispheres can write one answer and speak a different one.

The qualification 'most important components of speech handling' is significant because one of the essentials about the science of the brain is that the more we find out, the more complicated it is. So, for instance, the latest studies suggest that both sides of the brain deal with speech – but there is still a significant area on the left side not present on the right that is important to speech handling.

Similarly, as we saw in the introduction, until recently there was a tendency to label two different ways of thinking as 'left'- and 'right'-brain thinking. Left-brain thinking is the systematic, ordered, linear approach. It is the analytical approach of science and business.

Right-brain thinking, by contrast, is the artier, more holistic approach, dealing with spatial awareness, colour, music and the like. More research using fMRI scanners, which can monitor which areas of the brain are active during certain mental activities, shows that both sides of the brain are involved in both kinds of thinking, though one side is often more active. For convenience, the labels 'left'- and 'right'-brain thinking tend to be used for activities that have had long-term associations with that particular side.

The importance of left/right-brain thinking is that we tend to fall into one approach or the other, and the second mode largely switches off. This is a disadvantage if we are trying to be creative and come up with lots of ideas, as we want the whole brain in action. This is why many exercises to help get the brain into a creative thinking mode involve colour, music and spatial awareness, as the very act of sitting down and working on a problem or need for an idea tends to force us into a 'left-brain' mode of thinking.

If you have any doubt about the existence of the two modes, there is a simple and effective demonstration you can do, experimenting on your own brain. Visit the webpage universeinsideyou. com/experiment8.html and follow the instructions to try out the Stroop effect, which demonstrates this split in action.

LINKS:
• **Idea prompting** – page 172

Long-term memory

When we look back to our childhoods, memory has a trick to play on us – we remember things worse than they were. If you are over 30, you can't really remember what it was like to be a teenager.

It's frustrating that our children are convinced that we never had any of the experiences they endure; yet we feel we have plenty to offer. After all, we've been there. When they hit those traumatic teens we should be the obvious people to turn to – but in practice we rarely are. We lived through our teens and early twenties too. We can remember what it's like. And yet research shows that we don't remember correctly at all.

In an old but still valid study, a conscious effort was made to see how our memory of how we felt in the past is different from how we actually felt at the time. Back in 1944, around 500 American students, averaging nineteen years old, were asked to fill in a detailed personality questionnaire which explored their relationships with friends and family, uncovered how they felt about themselves and looked at the development of their social skills. In 1969, the same group were asked to take the test again. The outcome was very interesting. The basics measured by the test were surprisingly consistent for an individual between the two tests. However, the group was also asked to fill in the test the way they *thought* they would have done at age nineteen.

Now big discrepancies came out of the woodwork. The fortysomethings had a distinctly distorted view of what it was like to be a teenager. They remembered themselves as less confident, with badly developed social skills and generally unhappier than they actually were. When we look back on ourselves at this age, it seems that the angst and the failings come through much more strongly than they should. So the fact is, if you say 'I've been there; I remember what it's like,' to be a teenager, you are probably wrong. Not just, as they will assume, because you couldn't be like them because you didn't have their technology, but also because your memory is playing tricks.

The good news, should you feel sorry for teenagers, is that they really don't have it so bad as you remember.

On a more practical level, research from the US and China has underlined the importance of certain phases of sleep in consolidating memory. Specifically deep sleep, where we don't dream, seems to be important, as memories are effectively replayed and reinforced to ensure that they are retained. Although there is reasonable evidence that there is no need for us to get all our night's sleep in one block, this does suggest that very disturbed sleep or a sleepless night could make good memory formation difficult.

LINKS:
- **Brain-boosting habits** – page 156
- **Procedural memory** – page 186

M

· · · · · · · ·

Mind maps and note taking

Most children in schools are now taught mind maps as a matter of course – but even those who are taught them rarely use them enough.

There's a problem with the way most of us take notes. The way we do it doesn't fit particularly well with the way that our brains work. But there is a better way – mind mapping. Also called cognitive mapping, this is a way to capture information that fits better with the way our brains store information, and that makes it easier to retain information from our notes.

Making a mind map is simple. Turn a piece of paper sideways and draw a blob in the middle, labelling it with the topic you are making notes about. From that central blob you draw thick branches that correspond to the main aspects of the subject – then from each branch you draw twigs for the more detailed aspects (you can go down into more levels if you like). You write a few keywords on top of each branch or twig to cover the information. So, for instance, if I were drawing a mind map about mind maps, I might have major branches with items like 'What' and 'How'. Then the 'What' branch would have twigs describing what a mind map is, and the 'How' branch would have twigs describing how to draw one. A mind map about mind maps is shown overleaf.

There are several great things about using this format rather than traditional notes. Mind maps grow organically, structuring information as you go. The brain is better at remembering images than words – so mind maps can make use of illustrations to reinforce the message. And the structure helps you retain the information, particularly as it chains things together, just like the brain does.

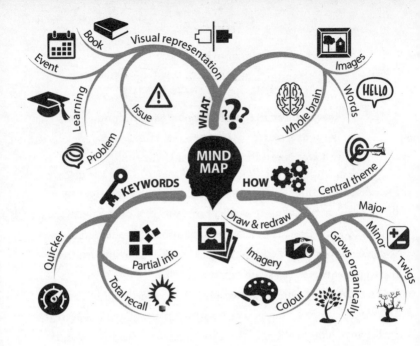

I really would recommend mind mapping when you have to take notes in meetings, when you are trying to structure something complex like a book, or when you are trying to absorb information from any source. Even those of us who are familiar with mind maps probably don't use them enough. There's a natural inertia that says 'it's too much trouble' – but it really isn't.

Mind maps often work best when drawn by hand, particularly as an aid to memory, but if you want a tidier version, perhaps to use as a handout, there is plenty of software available to produce them. These apps work particularly well on a tablet, where you can combine the clear typed text with the hand-drawn feeling of producing the branches and twigs with a flick of the finger.

LINKS:

- **Long-term memory** – page 177

Music

There's no doubt that many of us find music inspiring and relaxing. But can it also improve our intelligence? The evidence is scanty for listening but more interesting for performing.

If there is one thing that the commercial brain enhancement companies have jumped on, it is the impact of music on the brain, typified by the idea that playing Mozart to infants (even those still in the womb) can have a positive effect on their mental capacity. The popularity of this idea is due more to the convenience of the quick fix solution (just play a CD in the background) than any real confirmed benefit.

Unfortunately, there are two problems with the idea. One is that the study this theory is based on did not involve any research using babies at all. Working with infants is expensive for researchers and time-consuming to do, as you have to treat babies rather better than adult test subjects and because once you have done the initial work, you have to wait until the babies are five or six before you can make any sensible assessments of their mental capabilities. The actual research was a single small study (hence of little value in doing anything other than getting the media excited), using university students as subjects.

The second problem in making any conclusions of benefit is that all the study showed was a small increase in IQ immediately after listening to ten minutes of music – there was no attempt to measure whether there was any long-term impact or whether, as was more likely, this merely had an influence at that moment. And it just happened to be Mozart that was used. There is no evidence at all that it wouldn't have worked the same with anything from Albinoni to The Who.

What does seem to have a benefit is an aspect of music that takes considerably more work than just clicking on iTunes, and hence has received a lot less interest – a practical involvement in music.

Learning an instrument, for instance, gives a significant increase in IQ, though research has shown that the best improvement comes from singing lessons. Those receiving music training were compared with students receiving drama lessons and those receiving no lessons at all. Although the increase was relatively small, it was significant and spread across both IQ testing and general academic achievement. What isn't clear is whether the impact was from the actual lessons or from the discipline of practising and performing. It is entirely possible that there is no need for expensive lessons and that it is sufficient to try teaching yourself and taking part in a band, orchestra or choir.

However, like many such studies, particularly in the social sciences, we have to be aware of the possibility of a correlation that does not reflect causation. It is entirely possible that families who put their children in for music lessons are more likely to be generally supportive, and it could be the overall environment that provides the benefit, rather than the lessons.

LINKS:

- Causality and correlation – see **Paracetamol and childhood asthma** – page 313

O

· · · · · · · ·

Oxytocin

**You may have seen references to the 'love
hormone' oxytocin. But is there really
a chemical substance in the brain that
is responsible for romantic feelings?**

When talking about the limits of science, people often say that you
can't apply science to love or to the other feelings and insubstan-
tial aspects of life that are so important to being a human being.
However, most scientists would disagree. There are chemicals active
in the brain that influence the way we feel, both positively and negatively – we can't separate feelings, any more than we can memories,
from the electrochemical actions of the brain.

There are many different hormones and other compounds that
act in the brain, and to detail them all would require a (rather dull)
book in itself. But oxytocin is certainly one of the stars, with a range
of roles, cropping up in social behaviour, pair bonding, anxiety and
more. Like most such chemicals, oxytocin is released from certain
parts of the brain, moves with the blood and locks on to 'receptors',
specially shaped sites in other parts of the brain, where only specific chemicals can attach themselves. The receptors act like control
switches in the brain, increasing and decreasing certain activities
and indirectly resulting in feelings being expressed.

It is far too simplistic to call oxytocin the 'love hormone' and to
suggest that it makes us fall in love – but there seems little doubt that
it is involved in pair bonding and in social behaviour that gives us
a tendency to trust others and be attached to them. This role sometimes even works across species. When a human strokes a dog for
some time, the oxytocin levels shoot up in both of them.

So while there is no reason to think that a quick dose of oxytocin will act as a love potion, it does seem this complex hormone has a contributory role in the way our brains undergo emotional shifts. However much our feelings may seem detached from our physical bodies, and hence unable to be subjected to scientific analysis, it does seem that there is good evidence that compounds like oxytocin have a contributory role to play in what makes us the individuals we are.

LINKS:
- **Serotonin** – page 199

P

. . . .

Problem solving

We sometimes have a simplistic view that we only solve problems when something has gone wrong, but problem solving occurs in all kinds of human brain activity.

The most famous quotation when it comes to problems is: 'Houston, we have a problem.' (In fact, Jack Swigert, one of the astronauts on the ill-fated Apollo 13, actually said: 'Houston, we've had a problem here.') The word 'problem' immediately produces the image of some sort of mess that needs sorting out. But problem solving is a far wider activity than this and one for which the human brain is remarkably well suited.

Broadly speaking, we are problem solving whenever we think 'How do I do this?' – or rather whenever we act as if we have thought it, as much of our problem solving does not involve conscious thought, but makes use of, for instance, procedural memory. We are problem solving when we take on a puzzle, work out the murderer in a detective story or decide what to have for tea, just as much as we are when trying to fix an ailing space craft or a double booking in our social calendar.

Problem solving and generating ideas are two sides of the same coin. In solving a problem, you come up with an idea to provide that solution. When generating an idea, you are solving the problem that is the need for that idea.

Often problem solving is a trivial activity, something that is achieved without even thinking about it, or that is undertaken by making a simple decision (as in 'What should I have for tea?'). Humans have a tendency to make decisions more complex than they need to be. If there is a decision between alternatives that you are struggling to make, it can be a useful exercise to make the decision by tossing a coin. If you are happy with the outcome, then the chances are it really didn't matter which decision you made, so there was no point spending any further time on it. If you are unhappy with the outcome, you had already made a decision, but were not accepting it.

To solve more complex problems it can be helpful to use the kind of idea-generating techniques described in the **Idea prompting** section.

LINKS:
- **Idea prompting** – page 172
- **Procedural memory** – page 186

Procedural memory

We tend to think of memory as being about remembering facts and figures, but the oldest part of our memory in evolutionary terms concerns how to do things.

Human memory isn't a single store like a computer's memory – different parts of the brain are used to handle different types of memory. Right down at the bottom of the brain is an almost separate section known as the cerebellum. This is sometimes called the most 'primitive' part of the brain. This is misleading in the sense that our ancestors didn't start with just a cerebellum and gain the rest, but it is the part of the brain that is most similar to the same structure in fish, birds, reptiles and other mammals.

The positioning of the cerebellum is significant. Being closest to the spinal cord it can respond most quickly to an urgent need for a physical response, and it is responsible for a lot of our motor function. In effect, the cerebellum deals with 'how to do physical things' – which psychologists call procedural memory. When we first learn to do something, we tend to be very clumsy at it, because we are consciously thinking about how to do it. It was true when we learned to walk, but as we can't remember this, a better example is learning something like the basics of riding a bike, playing a musical instrument, typing on a keyboard or driving a car.

Think, for instance, of changing gear in a car. As a learner driver we have to remember to check which gear we are in, think to depress the clutch and ease up on the accelerator, shift the gear lever to the correct next position and reverse the clutch/accelerator combo. This takes so much concentration we often get it wrong, or have a tendency to drive into parked cars while we are doing it. Eventually, though, it becomes 'second nature'. This is because the memory of what to do has shifted from our information-based conscious memory to procedural memory in the cerebellum. We don't have to think about how to do it any more. We just decide what's wanted and the cerebellum handles the rest for us, without any conscious intervention.

The keyboard example is a good one to explore what is happening in a little more detail. Like many whose job involves using a computer all day, I can touch type. I was never taught, but simply picked it up as I went along. I don't look at the computer's keyboard as I type; I look at the screen and watch out for finger slips. Doing this, I can type quickly and effectively. But ask me where the B is on the keyboard and I can't tell you. I have never *learned* the layout of the keyboard in my 'facts and figures' high-level memory. (Okay, I know the top line is QWERTYUIOP, but that's all.) Yet despite not knowing where the B is, as soon as I want to write a B my finger goes there and presses it for me, using a totally different, unconscious form of memory.

LINKS:
* **Long-term memory** – page 177

Puzzles

We are often told that doing puzzles – from Sudoku to crosswords and visual mind games – are good for us because they 'give the brain a workout', but do they?

The good news is that puzzles aren't bad for us, so if you enjoy doing them, fill your boots. But think of them primarily as an entertainment, rather than a means of improving the performance of your brain.

Doing puzzles does indeed improve a mental capability, but that seems to be pretty well entirely limited to the ability to do puzzles. If you do puzzles, you get better at doing puzzles. But there has proved to be very little crossover discovered between puzzle expertise and other mental capabilities.

During the Second World War, the ability to solve crosswords was one of the tests used to select codebreakers for the UK's Bletchley Park secret establishment. Does this mean, then, that the spooks got it wrong? Not entirely. Someone who is extremely good at cracking crossword puzzles is demonstrating a skill in two things that are important in codebreaking – spotting patterns in lines of characters and deducing someone else's thinking from concealed clues.

However, it seems to be not that doing crosswords made these individuals better codebreakers, but rather that having a natural ability for crosswords also suggested an ability for codebreaking. It wasn't that doing the crosswords was training them as codebreakers, but that being extraordinarily good at crossword solving – better than most of us would be even with 30 or 40 years of such 'training' – was a reasonable flag to give someone more attention.

LINKS:
- **Brain training for fading capabilities** – page 162
- **Brain training for intelligence** – page 164

R

. . . .

Reading

**There's nothing an educationalist likes better
than to encourage us to read – and there is
evidence that houses with plenty of books tend to
produce children with better mental capabilities
– but what is cause and what is effect?**

We make a very natural assumption that there is a link between living in a house full of books (assuming they are read, rather than just bought by the yard as decoration) and improved mental capacities. Surely, reading 'expands the mind'. But what is the evidence?

The news stories certainly suggest that if this link exists, we have a problem. There are regular scares in the media about the way hardly anyone reads anymore and how it will bring the end of civilisation as we know it. We hear that 'only one in five parents easily find the opportunity to read to their children', 'only 40 per cent of England's ten year olds have a positive attitude to reading', and 'English 16 to 24 year olds were 22nd out of 24 countries measured for literacy by the OECD.'

On the other hand, reading is a more popular activity than gardening, going to the cinema, going to the theatre and concerts, or DIY – though inevitably it lags well behind TV viewing. But does reading do anything different to our brains from watching the television? We certainly regard reading as more socially desirable. And it's interesting that in a 2009 survey it was found that 61 per cent of British people have lied about reading a book they've never opened.

Apparently the book that was most lied about was George Orwell's *Nineteen Eighty-Four*, with a massive 42 per cent falsely claiming to have read it. As for reading versus TV, there are plenty of scare stories about the television, but in fact, on balance, it seems to be beneficial, provided we watch a range of programmes including news and documentaries rather than just soaps and reality TV. And because TV watching is a shared experience, it is more likely to produce discussion and debate than reading a book. However, a couple of areas where books do give an advantage over TV are in improved focus and concentration – it's much easier to skip around with the TV – and enhanced vocabulary and writing skills.

It is important, of course, to see if these benefits of reading actually carry through into everyday life. There is no doubt at all that *literacy* is a significant and important skill. Being able to read has a very clear and well-researched link to the ability to get a good job and to get on in life. And gaining an enjoyment of reading goes hand in hand with developing good literacy skills. But once we've got to that level, is there any further benefit?

There is good evidence that secondary school children who read for pleasure do better at school; though while we know the two go together, it doesn't mean that we know that the reading *causes* the better achievement. It could easily be that a home environment in which children have been encouraged to read for pleasure is also a home environment where children are given more help and encouragement to achieve at school. There's a similar unclear linkage for houses with books. Again, having lots of books correlates with better achievements, but both could be caused by a separate factor.

This doesn't mean, though, that there are no academic benefits from reading, over and above the pleasure gained from the activity. Reading has been shown to provide good inspiration in terms of academic direction, and can clearly add to general knowledge that can be used in an academic setting when reading non-fiction books, as well as a greater appreciation of the use of words and writing skills gained from fiction reading. There is also some evidence of reading being a good form of de-stressing, better than listening to music or going for a walk.

There has also been a study which showed that reading a 'gripping novel' (rather than literary fiction) produced improved connectivity in and stimulation of areas of the brain usually involved in actual activities. These effects lasted for at least five days, and the study was portrayed as showing that 'brain function was boosted by reading a novel'. However, it's worth bearing in mind that this was a single, relatively small study, and all that was discovered was that there were changes to the brain, not that they were necessarily beneficial.

One last specific – see more in **Brain training for fading capabilities** – is that reading is one of the activities that has been shown to help slow down mental decline as we get older. So even if it doesn't necessarily give the brain boost per se, it certainly is of benefit in keeping mentally fit and active – and is an activity we can all enjoy.

LINKS:
- **Brain training for fading capabilities** – page 162
- **Brain training for intelligence** – page 164
- Causality and correlation – see **Paracetamol and childhood asthma** – page 313

Remembering lists

Remembering long lists, whether it's a shopping
list or revision for an exam, can be a pain. But
there are tried and tested ways to make it easier.

We've all been there. You can't be bothered to write a shopping list because you only need half a dozen things. You pop to the shop, make the purchases and just as you get back to the house you realise that you forgot the most important thing. (It seems to get worse as you get older.) And yet at the same time you see people on the TV who can memorise whole pages from the phone book or a long sequence of cards without any problem. It's easy to assume this is down to some inherent capability – that they have some sort of special 'memory gene'. But in reality it's mostly a matter of using simple methods to trick the brain.

Almost all memory feats are based on the fact that human brains did not evolve to handle words and numbers. We are much better at storing and retrieving images and gossip than we are at tucking away facts or figures. So, to make it easier to remember, for example, that shopping list, all we have to do is to turn it into a dramatic, visual story.

If you are dealing with a shopping list, don't just remember the items, but make up a mental story where these items take part in a dramatic adventure. Make it is as colourful, visual, over the top and dramatic as you can and you will find it will stick very easily. A good way to get some practice is the childhood game 'In my bag I have …' where members of a group recite what's in a bag in turn, each listing everything in the bag, but adding one more item. Try to build a story as this game is played, reinforcing the story each time an item is added. It becomes very easy to remember 20 or 30 items flawlessly.

Going beyond a basic list, there is also good evidence that our ability to recall many things depends on our ability to make categories. If, for example, you ask someone to write down a list of all the animals they can think of, they will do significantly better if you first give them a set of categories, like farm animals, sea-life, local wild birds, zoo animals, etc. Our brains don't store things in straight lists, but rather as branching trees with links (see **Mind Maps**). If we can provide a good range of starting points, then we have a much better chance of tapping into these trees, compared with simply producing an unstructured list. This explains why mind maps make excellent ways to note and retain information.

If you want to memorise anything from a very long shopping list to the periodic table of the elements, don't try to take in the thing as a whole. Start with a set of categories that work well for you. These may be the obvious ones suggested by the type of information (the columns of the periodic table, for instance), or the categories could be more personal. So you might organise your shopping list by the different places you store things in the home, rather than the conventional divisions of the supermarket shelves.

If you are dealing with something that doesn't naturally break down into categories (though most do), it's always possible to impose artificial categories. Two good examples of this are to use a well-known place or journey. Each of a set of locations in your house, say, or familiar sites along a journey, is used as a mental repository. By placing mental imagery of whatever is to be remembered at this location, it becomes a special kind of categorisation. Place the thing you are trying to remember mentally at this location and use some kind of linkage (ideally dramatic, colourful or vulgar) to link the item to be memorised to the location. This technique has been popularised by the use of 'mind palaces' or 'memory palaces' in the TV show *Sherlock*, but the method goes all the way back to the Ancient Greeks.

LINKS:
- **Long-term memory** – page 177
- **Mind maps and note taking** – page 179

Remembering names

Most people claim to be poor at remembering names on a first meeting. It's not surprising – the structure of our memory makes them tricky to recall. But it's not a lost cause.

Probably the most frequently heard complaint about memory is 'I'm terrible at remembering names.' It's not surprising. Firstly, our system for interacting with the world relies far more on broad patterns than individual details. Identifying someone as a person is easy, but putting a name to which individual person is much harder. Then there's the matter of words being trickier than images. It's often said: 'I know the face, but I can't remember the name.' Words are a relatively late addition to the human armoury and not yet as well accommodated in memory as are images.

However, it is perfectly possible to meet anyone from a single individual to a whole collection of new people and, with a bit of effort, commit their names to memory. The simplest approach is repetition, a valuable throwback to the infant development of human brains, where repetition is all-important. Say a person's name aloud as often as you can at your first meeting and you are far more likely to recall it later.

To get a stronger anchoring of a name, the best approach is to make use of the strengths of memory and use them to tie a name to a person. Let's imagine I had just met a person called Marcus Archer and was trying to remember his name. The idea is to create a strong visual image – making it graphic, colourful and over the top – and use that to bring back the name. Some surnames, like Archer are easy because they come with visual imagery built in. But a name like 'Marcus' is more of a challenge.

With a name that doesn't have any visual connotations, use any other links that name may already have for you. Do you have a friend called Marcus? Is there a celebrity or a character in a book or TV show with that name? Or a historical character for whom you have a strong visual imagery? Make use of that. So, for instance, if I had a friend called Marcus, I might imagine my friend, dressed as an archer, shooting enormous rubber arrows at the person whose name I was trying to remember.

Some names are even more challenging. If it's long, you might have to break it down into syllables. Look for rhymes, contractions and associations – anything that will help pin the name down. The first person I ever tried this technique with was called Ann Hibble. I imagined a huge purple hippo bursting out of the floor of Boots (which is where she served me) and nibbling her toes. 'An Hippo Nibble' – Ann Hibble. It's very silly, but this was a person I met just once over twenty years ago, and I still remember the name.

One final essential with this or any attempt to get something into long-term memory is reinforcement. Come back and recreate the imagery regularly until the image eventually becomes unimportant because you *know* that name.

LINKS:

- **Baby brains** – page 153
- **Long-term memory** – page 177

Remembering numbers

There was a time when we all just about managed to remember telephone numbers, but mobile phones have pretty much killed this ability. If you need to remember a number, though, there's an easy way to do it.

In the old days you might have had to remember a telephone number like 0208-562-4169. It's tricky to remember a whole telephone number, as short-term memory can only cope with around seven items at a time. What is cunningly done is to break the number up into chunks, as we can often take in a chunk as a single item. In this case, for instance, I would have known that 0208 was the code for outer London, so I could think of that as a single item.

Mobile phone numbers are potentially harder, because there is no convention for breaking them into chunks, though this does at least allow for some flexibility to match any repetition or patterns in the number. For instance, my mobile number has the digits 8080 in the middle, so when I give out my number I always break it down to the first bit, 8080 and the bit after.

Chunking helps enable you to briefly store a number in short-term memory, but you may want to hang on to a number for longer than the few seconds that short-term memory keeps a grip on information. In that case, it helps to have a technique to encourage information into long-term memory. Just as with names, this is best done by using imagery and, like lists, a story is a great way to get that imagery into a sequence. The problem is that numbers don't lend themselves to stories.

To get around this, most memory experts use a simple rhyming scheme, replacing numbers with words. A typical set would be:

1 – Gun

2 – Shoe

3 – Tree

4 – Door

5 – Hive

6 – Sticks

7 – Heaven

8 –Weight

9 – Line

(1)0 – Hen

If you have a good mental image of a hero, you could make that last one zero – hero instead, but otherwise, ten/hen makes a good substitute.

By simple repetition, get these rhyming words into memory so that given a number (6) you can instantly come up with the word (sticks) – and vice versa. Then it's simply a case of building a story, as usual making it dramatic and outrageous, featuring the items in the order required to spell out the number.

To begin with, you will remember the number by working through the story, but before long the number should pop into your mind automatically, properly located in long-term memory.

LINKS:
- **Long-term memory** – page 177
- **Remembering lists** – page 192
- **Remembering names** – page 195
- **Short-term memory** – page 200

S

* * * * * * * *

Serotonin

**Like oxytocin, another complex chemical that has
a powerful impact on the brain is serotonin.**

Serotonin is a neurotransmitter, a hormone that binds on to
receptors, having a significant influence on the operation of the
brain. Surprisingly, though, the majority of the serotonin in your
body never gets anywhere near your brain as it spends its time
moderating the function of the gut. At a basic animal level, this is
clearly linked to perceptions of food availability, but in the human
brain it has come also to have an influence on the mood of the
individual.

For this reason, many antidepressants act by keeping serotonin
levels up, in the hope of improving the patient's mood. (Psychedelic
drugs from LSD to the psilocybin in magic mushrooms tend to
block the receptors that serotonin locks on to, producing dramatic
mood swings.)

Serotonin is thought to be responsible for physical exercise's
positive effect on mental state, as having a workout can help increase
serotonin levels. Other possible approaches to improve serotonin
levels without resorting to medication are bright light and some
aspects of diet.

Special sun-like bright lights are often prescribed for Seasonal
Affective Disorder (SAD), the condition of feeling low or depressed
in the winter. It is possible, though not proven, that the lights help by
pushing up serotonin levels. Most of us certainly feel more positive
on a bright, sunny day than a dull, cloudy one; so where there is an
opportunity to get out in the sunshine (with suitable UV protection)
or have a spell with bright lights, it may be beneficial.

When it comes to diet, there are some myths doing the rounds. You may have seen that high-protein foods like turkey or foods rich in serotonin like bananas will push up serotonin levels and improve your mood. This won't work, as serotonin does not cross the blood-brain barrier that protects your brain from infection, so consuming food with serotonin in it doesn't increase levels in the brain. However, it is just possible that some foods that are rich in tryptophan – a protein that does seem to trigger an increase in internal serotonin levels – could be beneficial for the mood. Chickpeas are a good source. If there is anything in this theory, it is best to avoid a diet heavily based on sweetcorn (maize), as this is unusually low in tryptophan.

LINKS:

* **Sunburn** – page 320

Short-term memory

The short-term memory is where we hold things briefly before moving on. It's something we are hardly aware of, yet it is very important for being able to function well mentally.

In effect, we have two types of short-term memory. Basic short-term memory is where we hold a number, say between reading it and punching it into a phone. This part of your memory is very limited. Most of us can only hold around seven items – which could be a number or chunk of numbers, a name or an image, for instance – and even the best short-term memories typically only manage around nine items.

Of itself, the size of our short-term memory doesn't make a huge contribution to our cognitive ability, but a related kind of memory, working memory, does seem to be relevant. Working memory is where you manipulate items that you are holding in short-term memory and is necessary for a wide range of mental activities. You would need to use working memory, for instance, to swap two digits in a list of numbers, to do mental arithmetic or to sort words into alphabetical order.

While there is not a huge amount of benefit from increasing short-term memory capacity, there is reasonably good evidence that we can get some impact by improving our working memory capabilities using the kind of exercises described in the **Brain training for intelligence** section. These exercises almost certainly improve the ability to take IQ tests, although it is less definitively certain that they improve the kind of cognitive ability that helps with actual life.

LINKS:
* **Brain training for intelligence** – page 164

Software

**Computer software now routinely provides a
huge support to our mental capabilities.**

As is made clear in the **Brain training for intelligence** section, while it is possible to train the brain to make better use of working memory (which can be seen in some respects as boosting intelligence), so-called 'brain training' software rarely improves things, though like all mental activity, it can help prevent deterioration. However, in a different sense, through the advent of personal computers and the internet, software has made a huge impact on the brain as an external support and extension network.

It would be pointless to try to cover every possible aspect of software that extends brain function, but here are three good examples.

The internet. Both in terms of communication and access to information, the internet has transformed the way we interface with the world, at a level that is primarily the concern of the brain. Our brains seem to be coping with this very well – unlike some post-evolutionary developments, we have taken to the internet like ducks to water. But our education system lags behind, still expecting us to retain large amounts of unnecessary information. The education system will have caught up only when we provide students with access to the internet during exams. If you think this means they will just look up the answers to questions, then the questions you are thinking of are inadequate. We should be testing how people understand and use facts – access to the facts themselves is too trivial to test once you have the internet. (Though assessing the quality of 'facts' found on the internet is a valuable skill.)

Note taking. Although a lot of information is readily available online, it doesn't provide access to a record of your own thoughts, which might be anything from your notes from a book or a meeting, to ideas, a business plan or shopping list. Note software (like Evernote) makes it possible to make notes wherever you are on any device and to have it accessible anywhere else. For example, I have Evernote on my phone and can use this to access essential information wherever I am. A while ago, I went into town and needed some of that information – only I had left my phone at home. No problem. I just stopped by the town centre library, got on an internet-connected computer and could check my notes online.

Calendars. Or schedules or diaries. We have long recognised that our brains need assistance with scheduling and used to carry a paper diary to deal with it. Unfortunately, that wasn't available all the time – but like notes, a modern software-based diary can be accessible wherever you are via phone, tablet or computer. What's more, you can link diaries so that, for instance, you know what's in your partner's diary as well and don't double book.

LINKS:
- **Brain training for intelligence** – page 164
- **Mind maps and note taking** – page 179
- Working memory – see **Short-term memory** – page 200

T

· · · · · · · ·

TV watching

**There are regular scare stories about the dangers
of watching too much TV and the negative
effects it has on the brain – but are they true?**

As is explained in the **Baby brains** section, there is evidence that for
the first two to three years, TV is not a great medium to leave a baby
with unsupported, because at that stage so much is dependent on
feedback. Babies are copying and want us to respond. When they
copy a person on the TV, the person doesn't respond. But once we're
past this stage, is television really an 'idiot box'?

Early doubts were based on a snobbish certainty that TV was
common, lowbrow entertainment, while reading a good book (as
long as it was one of the classics) was educational and uplifting.
This still lingers on to some degree, but there do seem to be some
negatives to TV watching, particularly for children, that require
addressing.

The first is not brain related, but rather concerns exercise. There
is no doubt that spending hours slumped in front of the TV is not
great for your physical health. The concept of the 'couch potato'
isn't entirely a caricature. If you spend all your spare time in front
of the TV, you are not getting enough exercise. (In principle, this is
a problem with books too.)

It is also possible that too much exposure to TV, particularly in
the early years, could increase levels of the neurotransmitter dopamine. It has been suggested that this is particularly the case with
modern TV editing styles, which use lots of fast cuts, rather than
staying with a more natural, steady view. While there is no hard and
fast evidence, there is a concern that having high dopamine levels

for long periods could damage the brain's reward mechanisms that enable us to achieve and to take interest in the world. This is a concern rather than something with good evidence to back it up, but it may be worth restricting TV viewing, particularly for pre-teen children.

On the plus side, there is no doubt that TV entertains and informs, and that for many older people it is a lifeline that makes what could be a lonely existence more bearable.

What seems to come out is that for adults there is no real danger, but we need to ensure that we are doing other things as well – it's the mix that matters for the adult brain. For children we should be a little more wary and ensure that there isn't too much uncontrolled TV watching. Keeping TVs out of pre-teen children's bedrooms seems an obvious step.

LINKS:
- **Baby brains** – page 153

V

· · · · · · · ·

Video games

**We hear everything from suggestions that
video games can train our brain to fears
that they are turning us into 'computer
zombies'. What is the reality?**

As covered in the **Brain training** sections, there is some evidence
that playing games that require thinking skills helps ageing people
keep fading capabilities in check, but very little evidence that games
improve intelligence. Over and above this, the only positive impact
from computer games other than their entertainment value is that
they can enhance hand-eye coordination.

What, though, of the negatives? Baroness Susan Greenfield is
professor of pharmacology at the University of Oxford, but spends
a lot of time talking about the dangers to the brain from the internet and playing video games. (Ironically, she herself marketed an
expensive 'brain training' game.)

She has claimed that computer games can cause temporary
dementia in children, telling a conference that an addiction to this
technology could disable connections in the brain, literally blowing
their minds. On many other occasions, she has commented on the
dangers to the brain, particularly children's brains, of video games,
computers and the internet.

However, as has been pointed out by *Bad Science* author Ben
Goldacre, there is something strange about Baroness Greenfield's
claims – they are always made in the popular media. The usual
mechanism for a scientist to present a theory is through a peer-
reviewed scientific paper, which will be assessed by experts,
but, as Goldacre commented back in 2011, 'Why in over 5 years

appearing in the media raising these grave worries has Professor Greenfield never simply published the claims in an academic paper?'

As yet there is no good evidence for damage to the brain from excessive playing of video games. As mentioned in the **TV watching** section, there are sensible health concerns from spending too much time immobile in front of a screen, but this is quite different from being damaged by the activity. It's also the case that some modern games involve very graphic violence. This feels like it should be bad for the player, but all the evidence is that, provided the player is old enough to have a clear distinction between fact and fiction, the game is accepted as a story, rather than reality.

We see reports of people who go on to commit mass murder who have been enthusiastic players of violent video games, but as yet there is no evidence that playing the game was the cause of the real-world violence. The vast majority of game players don't commit violent acts as a result of the games, and it would be equally possible (though equally without evidence) to argue that playing such games could prevent violence, as it gives the player a chance to take out their aggression in a harmless forum.

As with all such entertainment, the essential seems to be variety. I played a lot of computer games in my twenties, and can testify to the addictive nature of some games that kept me playing long into the night – but I have also read books long into the night on other occasions. As long as the game play isn't the only interest, it seems relatively harmless.

LINKS:

- **Brain training for fading capabilities** – page 162
- **Brain training for intelligence** – page 164
- **TV watching** – page 204
- **Violence on TV and in film** – page 208

Violence on TV and in film

The debate has raged for years as to whether or not too much violence on TV and in film has a negative impact on our brains. What is the current view?

Much like the video games debate, there has been much discussion of the effect that watching violence on TV and in films has on the human brain, particularly in young people. In fact, a variant of this concern goes all the way back to the Ancient Greeks, when Plato complained about the bad influence that watching plays was having on youths.

There certainly have been hundreds of studies concluding that there is some linkage between watching violence on the screen and actual aggressive or violent behaviour. However, these studies have to be taken with a pinch of salt. This is a very difficult area to control for other possible causes, and the vast majority of the studies have been so flawed that they can provide no meaningful information. One interesting observation is that during a period when TV and film featured increasing levels of violence, the levels of youth violence consistently fell – hardly suggestive of a strong causal link.

Like many statistics outside of the laboratory, it is difficult to be sure why these levels fell. One study has suggested that levels of violence in young people might be linked to the level of atmospheric lead, which rose with the numbers of cars on the roads, then fell dramatically when unleaded fuel was introduced. Others have suggested that the drop in violence was linked to the wider availability of abortions a generation earlier, leading to fewer unwanted pregnancies and fewer children being brought up in social conditions that encourage disruptive behaviour. But the statistics certainly don't help the suggestion that TV watching encouraged violence. This is a good example of data where it is difficult to decide between causality and correlation.

Without good quality scientific evidence, all we can do is speculate that it seems reasonable to keep exposure to screen violence to a minimum in pre-teen children. When our children were around ten, we were amazed when a straw poll was taken at their school of how many of their peers watched the TV programme *Buffy the Vampire Slayer*. The majority claimed they did. While some children would say this to show off, the chances are that a good number of them actually did – but this was a show that is rated 15–18 for violence and sexual themes.

I don't say this in criticism of *Buffy*, which was an excellent series, but rather as an emphasis for the need to provide some control in what younger children watch as a precaution. The two biggest difficulties, assuming parents don't just use the TV as a babysitter, are TVs in children's bedrooms and watching with older siblings (this proved a particular problem in the *Buffy* example). We still don't know for certain that there is any benefit from restricting access to violent TV and film for pre-teens, but it is a precaution that seems worth taking.

LINKS:

- Causality and correlation – see **Paracetamol and childhood asthma** – page 313
- **TV watching** – page 204
- **Video games** – page 206

PSYCHOLOGY

Although psychology is all about how we use our brains, it deserves a separate section that isn't so much about how to improve (or avoid damaging) our brains, but rather how the brain can be manipulated. In a sense, the **Brain** section is the technical manual, but **Psychology** gives us the user guide.

There has always been a form of folk psychology – an understanding of 'how people tick' – and as a result, an interest in how to manipulate them. This came to the fore with the rise of marketing and advertising in business. Pretty much in parallel, academics began to look at the science of the way we behave and the realities behind the guesswork.

The results can be quite surprising. We aren't always the rational beings that we might expect ourselves to be, and the benefit of knowing a little more about psychology with the *Science for Life* approach is both to understand your own behaviour, but also to be aware of when others are trying to manipulate you, which they inevitably are. That way you can make decisions based on fact and what's best for you, rather than making the decisions those would-be influencers would like you to make.

A

. . . .

Advertising

**If psychology is the study of the mind's
function and human behaviour, advertising
is its most powerful applied discipline. It
helps to know a little psychology to be aware
of what advertisers try to do to you.**

Advertisers are out to persuade you. To buy a particular product.
To vote for a particular party. In the end, their aim is to get you to
behave in a certain way – and they have many tools at their disposal.
Here are a few of their key methods, to help you spot them and
make the decision you want, rather than the one *they* want.

- **Information**. This is the neutral face of advertising – the one
  they want you always to perceive. Sometimes factual information is enough. If you like a particular band, all the advertising
  has to do is tell you how to buy tickets and attend the gig. If you
  like a particular make of chocolate, you would like to know they
  have a new product out to give it a try. But very little advertising
  relies solely on information. For example, there is the appeal of
  novelty.

- **Novelty**. To overcome the power of habit, we need something
  new to make us consider switching. This can be as simple as
  a new product on the market, an improvement to an existing
  product – how many times have you seen advertisers use the
  claim 'new and improved!'? – or the provision of a feature that
  we have not experienced before, but are assured we'll like.

- **Comparison**. We like to make comparisons to be assured we are getting the best deal, and often advertising uses this. The simple form is to compare with a competitive product and show why this product is better – but advertisers have other tricks up their sleeve. These days, many magazines are available online or in print. It's not unheard of for an advert for a magazine subscription to offer the online version at, say, £50 for a year, the print version at £100 a year, or the print-plus-online for £95 a year. You may well be happy with the cheaper online-only version. But it is difficult to resist the bargain you appear to be getting by paying less than the print version for the print-plus-online double. Similarly, when I recently bought a car, I was persuaded to pay more than I intended because the model I bought was cheaper than the model below it in the range, thanks to a special offer. Comparisons can distract us from the actual cost because of the apparent bargain. We assess things relative to others, rather than in isolation, making comparison a powerful tool.

- **Range**. Rather like comparison, advertising uses ranges to push the items they particularly want to sell. Most people feel that buying the cheapest wine at a restaurant makes them look penny-pinching, so will go for the next wine up in price. Most restaurants make this 'second worst' wine the one they can squeeze the most profit from.

- **Reduction**. You might think price reductions are just that – but actually they cunningly produce a comparison that doesn't exist. If you see something reduced from £50 to £10 it is clearly a bargain, even if it is something that you would never think of paying £50 for, and really only want to pay £5 for. Although advertising regulations try to prevent it, advertisers and marketers can almost always find a way to make that 'original price' more than they would ever expect to sell it for.

LINKS:

- **Habits** – page 226

Air versus road

One of the essentials of understanding human psychology is pinning down when and where we behave irrationally – and rarely is this more obvious than in the fear of flying.

Many of us dislike or are scared of flying. This is an interesting aspect of psychology because it is both sensible and irrational at the same time. It is sensible because it is not exactly a natural situation to be stuck in a plastic and metal tube, closer than is comfortable to hundreds of strangers, several miles up in the air with no visible means of support. But it is also irrational because flying is a very safe way to travel.

The chances of being killed in a plane crash in any year are about one in 125 million passenger journeys. This makes the risk of a particular journey about three times safer than a year's worth of train journeys and twelve times safer than a year's worth of car journeys. Bear in mind, though, that on average we make significantly more rail trips and vastly more car journeys than flights – you are around ten times more likely to die on a particular plane journey than a particular car journey. But even so, the risk is very low – you are more likely to be killed by lightning in any particular year than in a plane crash, and it's riskier to be in the typical workplace than on a plane.

So why is flying so scary? Apart from that unnatural environment, a major factor is the amount of media coverage that is given to plane crashes. When an airliner comes down it is a major news story. Accidents on the road are not, unless they have many casualties. Typically no more than 2,000 people a year die in crashes of large planes worldwide, compared with around 1.25 million on the roads. (A high proportion of these are in developing countries where there are fewer road safety regulations.) But because we see plane crashes splashed across the media, we give them more mental weighting than they deserve.

Not only do we over-weight things that we hear about more in the media, we also give things more weight if they have a close connection. So, for instance, we are much more upset about 50 people being killed in our own country than 500 killed in a distant country with which we have no connection. This can also come through in lessons learned. All young people know that drink driving is dangerous. But they still often take the risk, because when young, humans are programmed to take greater risks and to assume that the worst won't happen to them. Plenty of studies have demonstrated that adolescents have an increased need for novelty and sensation-seeking, and this coincides with a time when their ability to regulate their own behaviour is not fully developed. Safety campaigns make very little difference to that behaviour – but research has shown that if one of their circle of acquaintances is killed or seriously injured in an accident caused by drink driving, there will be, at least temporarily, a change in attitude.

The strong weighting given to information that we are close to can be very misleading. We see it when people try to justify, for instance, smoking, where there are well-established risks, by saying: 'My uncle smoked 40 a day all his life and he lived to 96.' Such an argument over-weights a single piece of information because it is close to the person with the opinion, where what happened to a single smoker has no significance in the overall picture of many millions suffering smoking-related illnesses. Anecdotes don't get us to the truth because in something like smoking survival rates, we need to know averages, not just what happened to one individual.

LINKS:
•   **Risk** – page 235

Assertiveness

Discovering the distinction between being assertive and being aggressive can be the key to getting the best from a difficult situation.

Just imagine the situation. You arrive at an airport and your bags don't turn up. So you head off to the airline desk in a fury and give them a good telling off. It might make you feel momentarily better – but it won't help get your bags back. In fact, worse, it is well researched that taking an aggressive stance is more likely to turn someone against you and reduce the amount of assistance you get.

The same goes for any situation where you are at the mercy of someone else, frustrated and technically in the right, but unable to force the outcome you want – when, for instance, you take something back to a shop and they won't allow you a refund or a replacement. It might feel that if you don't go in all guns blazing you will be perceived as weak and will be trampled over, but the fact is that most employees have leeway (and if they don't, they will have a manager who does) – and how you act makes all the difference as to whether or not they will exercise that leeway for your benefit.

As you approach the desk or counter (it's far better to do this face to face than over the phone, though it isn't always an option), try to put yourself in the position of the person you are about to deal with. Bear in mind that, though their company may have done you wrong, they personally are not responsible, and will feel hard done by if you take out your frustration on them.

When you get there, smile, and try to smile regularly through the conversation. People can't help but respond better and more constructively to someone who is smiling. They will be more inclined to go out of their way to help. But avoid a permanent manic grin, as this can appear more scary than positive. When you speak, keep your voice controlled. Don't shout. Speak nicely and calmly. If it feels like this is not going to happen, take a couple of slow, deep breaths. Never lose it vocally, however much you are seething inside.

Connected to that voice control, don't allow anger to take over. Being assertive means that you are taking control, not giving in, but equally not being aggressive and unpleasant. Try to make your body language open and friendly. Don't push your face into theirs or wave your hands towards them. Keep restating your case – not mechanically, but calmly putting your position without showing any sign of backing off.

Keep your complaint as focused as you can. Don't tell them lots of irrelevant history and context. Simply state what has gone wrong and what you would like done about it. So, for instance, you might say: 'I bought this phone and it won't take a charge. Could you replace it, please?' But don't tell them you bought it three weeks ago from a guy with red hair, and it was very busy in the shop, and you intended it as a present for your auntie, and all the rest.

When you are asking for action, make sure you *are* asking, not ordering. Make it a polite request for assistance, which makes them the bad guy if they don't help, rather than a demand for action, which makes you a bully. You may well be turned down. Don't give in, or start ranting now. Make it clear with your body language that you are going nowhere, and keep up the pressure. If after several attempts they are showing no sign of taking action, say something like: 'I appreciate you are doing all you can, but I really need to get this sorted out. Is there a manager I can speak to or someone else who can help?'

There has been very little research in this area – most work on assertiveness has explored assertiveness within an organisation, rather than the reaction of staff to assertive customers. But my own personal, unstructured research while working with customer-facing staff in an airline, and running a PC support department, shows that assertiveness wins over aggression every time. Of course, being as assertive as you like doesn't guarantee a result. If, for instance, you turn up for a flight without your passport, it really doesn't matter how you ask to be allowed on – you won't be. But there are plenty of circumstances where there is some opportunity to get movement, and assertiveness is the way to achieve it.

B

• • • • • • • •

Bargaining

Don't assume you have to pay what's on the price tag. A surprising range of stores and tradespeople are open to bargaining.

When research has been done on shops and tradespeople's attitude to bargaining, the outcome is usually quite a surprise. The vast majority will accept some haggling rather than forcing you to accept the price on the tag.

Whether you are having work done in your house or buying a new tyre for your car, never assume that the first price you are quoted is the best one. There are a number of tactics that can be employed. Say that you have seen a better price elsewhere (this is best done where you actually *have* seen a better price, but that isn't essential). Say that you really hadn't expected to pay that much, and it wouldn't be practical. Then either say something like: 'Could you make it more like £X?' or 'Realistically, what's your best price?' The latter will probably not get you the best result, but it will conclude the transaction quicker if you find haggling embarrassing.

A fear of haggling seems to be a trait in some cultures – it feels that bargaining on the price is not appropriate behaviour. And yet, why pay more than you need to if you can agree a price that both of you are happy with? Most people find that with a bit of practice they begin to enjoy the process to such an extent that, for instance, if they turn up at a car dealer who refuses to haggle it spoils the whole process.

If you are offered a relatively small discount, or it's a big-ticket item like a house or a car, don't be afraid to go back several times. The opening price offer may well still have some leeway for bargaining. After things have shifted two or three times might be the best point to bring out the line: 'Okay, we're getting there. But what's your absolute best price?'

What is particularly surprising is that even large chains will respond to haggling. Insurance companies, phone companies and car breakdown services, for instance, will almost all come up with contract renewal offers which will then be dropped in price if you threaten to go elsewhere. Never take the first offer on the table.

Research has shown that at least 30 per cent of supermarkets will also haggle – perhaps the last companies we would expect to do so. You will probably have to ask to speak to a manager, and ideally would be buying a large amount of items or a large-ticket item like a TV, but apparently even some of these cost-cutting giants would rather reduce their margin than lose the sale.

In the end, the threat to walk away is your most powerful weapon. It's the reason that your mobile company or insurance firm will almost always move on price, and the same applies to everyday purchases. I learned this in a Croatian market as a teen-ager. I wanted to buy a necklace, but the price was far too high, so I started to walk away, but was given a lecture by the stallholder on the importance of haggling. In the end, I got the necklace for a third of the asking price.

C

· · · · · · · ·

Clusters

When things happen at random, they aren't spread out evenly, but occur in clusters. This can make it seem that there is a cause that doesn't really exist.

One of the biggest problems psychology throws at us, and one of the most important concepts to understand when trying to get a handle on how risk impacts your life, is clusters. These are a simple phenomenon of randomness that fool us into thinking something has a cause.

Imagine, for instance, that there is a cluster of outbreaks of cancer in a particular location. A natural inclination might be to look for a local cause for those outbreaks. And that is important for the experts to do – but only once they have eliminated the chance of a cluster occurring.

Imagine taking a tin full of ball bearings and dropping them on the floor. It would be a huge surprise if they all landed in a neatly spaced grid, each about the same distance from the others. Instead, we expect them to be more uneven. In some places there will be gaps, and in others there will be groups of ball bearings close together – clusters. That is what randomness looks like.

So when we see a cluster of cancer cases (or whatever) in a particular location, before blaming it on a local cause, say, a phone mast, we first need to see if this kind of cluster is exactly what we would expect without a cause. To do that, a statistician will need to look at the overall number of cases across the country and how they are distributed. There are good mathematical tests that show what sort of clusters we can expect, and whether a particular cluster is likely to be a result of random chance or due to some local cause.

To have an instant knee-jerk reaction, saying there must be a cause, is simply superstition. In the old days, they would have blamed the cluster on a local witch – now we resort to modern bogeymen like phone masts, high voltage power lines and nuclear power stations. (Interestingly, we rarely blame a cluster on a church or pub, yet there will almost certainly be one near many clusters.)

We also need to be careful how clusters are represented, as there are ways of presenting information that make them sound worse than they really are. This can be done accidentally or – all too often – with the intention of misleading the listener. (Politicians are always doing this.) Say, for instance, that you were told that in a particular company, 40 per cent of sick days are taken adjacent to a weekend. Surely this is evidence of people fiddling the system? No, it is just natural clustering. Forty per cent of the working week (Monday and Friday) is adjacent to the weekend. But we don't normally think of it that way, so it's easy to be misled.

The essence, when a coincidence occurs, is not to assume that there has to be a cause, nor to leap to conclusions as to what that cause might be.

LINKS:

- Causality and correlation – see **Paracetamol and childhood asthma** – page 313

F

• • • • • • •

Free things

There is something very special about getting something for free. So much so that we have to exert a real effort to avoid being sucked in by this most powerful of psychological tools.

A little while ago, the supermarket Waitrose made an apparently silly decision. They introduced a loyalty card, but rather than the usual approach of this giving you, say, 1p in the pound off your shopping, it meant that once a day you could get a free cup of tea or coffee in the Waitrose in-store café. Surely the company had gone mad? It would mean freeloaders piling into the shop day after day for their freebies, slashing company profits. And yet Waitrose seems to be doing very well on the arrangement – because they understood the power of 'free'.

The fact is, we don't respond well to a 1 per cent discount. It sounds too silly and small – we aren't prepared to do the sums and add it up over the year to see how much it benefits us. But a free coffee or tea is tangible, is delivered right away and has multiple benefits for the store.

First, we are likely to visit more often. It may be, especially if we live near a store, that some of those visits are literally just to get a free drink. But getting us in more often means that we have to pass through the store to get to the café more often – and could be tempted to make a purchase. At the café there is another insidious benefit for Waitrose. Because we are getting a free drink, buying a cake to go with it doesn't seem much of an extravagance. So we may well end up spending a couple of pounds on something to eat in exchange for getting a free coffee that perhaps costs Waitrose 10p.

There's more too. Supermarkets are often grim places that you visit with gritted teeth and stay in as short a time as possible. But now, Waitrose feels like a pleasant place to go – they give you free drinks! It changes your relationship with the store. A 1 per cent discount might seem the obvious way to go to an accountant, but my suspicion is that psychologists would recommend the free drink every time.

A more common example of the power of 'free' is the offer 'buy one, get one free' or 'buy two, get the third free'. Used appropriately, this is a genuine bargain. If, for instance, the price hasn't been inflated, and this is a product that has a long shelf life, which you will continue using over time, it is well worth stocking up. (This is true of any substantial discount.) At the time of writing, the best interest rate available on a UK savings account is 3 per cent a year. Imagine there is a product that I use that costs £10 a time, but it is currently being sold at £5. If I spend £50 on buying a year's supply in one go, I have got a 100 per cent return on my investment over the year, saving £50 on the full price. (This assumes that the product returns to full price in a reasonable timescale.)

However, all too often we buy things we don't particularly want – and won't consume in those quantities – just because they are 'buy one get one free'. A powerful example is soft fruit, often sold this way by supermarkets. Half the time, that second punnet of strawberries will be thrown away because they go off before you have a chance to eat them. But the supermarket doesn't care. You were tipped over the balance of 'should I buy strawberries?' – rarely a real need – by that word 'free'.

Probably the most important response to 'free' is to establish just what benefit you will get from the free item. Once we get over our childhood obsession with a gift (take a look and see how many children's magazines have a free toy on the cover), it's important to make sure that a free item has some genuine benefit before you allow it to sway your decision.

A second consideration to make sure you are taking the *Science for Life* approach is to add up any additional expenditure that comes along with the 'free' item. This is pretty obvious with the 'buy two, get the third free' offer (a dubious benefit if you only wanted one), but is more subtle in some cases. Do you have to fill in a questionnaire or wait in a long queue to get the free item? Costs aren't all about money. By all means, queue for a free item that you really want – but if you don't, think what sort of value you would put on that time and whether the item is really worth it.

LINKS:

- **Scale and understanding numbers** – page 239

H

. . . .

Habits

There are good reasons why habits exist. But in our modern world we need to question why we go along with them too.

It's a bit of an insult to call someone a 'creature of habit'. It suggests that they are unadventurous, taking the safe route, never getting far. But in fact, every human being is a creature of habit.

We can't help it, because of our deep dependence on patterns. The world is far too complex for us to have mental 'instructions' on how to deal with every possible variant of every object; so wherever possible we use patterns to label something and dictate how to react to it. We don't have to learn how to deal with every door – we learn the basic patterns of how doors work and it gets us through practically all of our interactions with doors … until someone puts a pull handle on a door that has to be pushed. (I've seen this, and it's fascinating to watch person after person come up to the door, try to pull it open, get nowhere and eventually push it.)

In essence, a habit is a behavioural pattern. We find an approach that works and that becomes our standard way of doing something. Even if there's something better that is easily accessible, we usually stick with the habitual. So, for instance, most people hate using a self-checkout at the supermarket when they first come across it. But if they use one several times and it works without a problem, it becomes the standard. They begin to realise how poor the traditional checkout is for a small basket of items. They have invested some effort in developing the self-checkout habit, and now it becomes the norm.

As someone who mostly works from home, I am shocked by

the number of friends that work in an office who casually pay £3 to £4 – often more than once a day – for a takeaway cup of coffee. This has become habitual to them. It's the norm. Psychologists would say it has become an anchor. I do visit coffee shops, but rarely for a takeaway. While it still seems an exorbitant amount to pay for a coffee, I can see it's worthwhile for somewhere comfortable and warm to sit for half an hour with free Wi-Fi.

It's a good exercise to see what your habits are, and to do a reverse scaling exercise. If, for instance, you visit a coffee shop several times a day, you could well be spending over £1,000 a year on coffee. Think what else you could do with that money. The same goes for smoking. If you are a smoker, an individual packet might not seem like a horrendous expenditure, but think what you could do with the money you spend in a year. (Another example is if you drink bottled water.)

Unless they are physically harmful, like smoking, there is no particular reason to break habits that you enjoy and where there isn't anything you would rather do with the money. But it is well worth seeing if there are some situations where the habit is hiding the fact that you would get far more out of that time and money used a different way.

LINKS:

- **Scale and understanding numbers** – page 239

O

· · · · · · · ·

Overvaluing things
we own

**When we own something it takes on a special
value to us. It becomes part of what we are
– and as such we tend to overvalue it.**

We sold a house once when prices were falling. It went for a lot less than we originally asked for it, and we felt hard done by. It was ours and it was precious. It's easy to forget that something you want to sell is only worth what other people will pay for it – and once we do own something, we mentally raise its value significantly.

It seems that we give more weight to how much we will suffer the loss of something that we already have than we do to the benefits of what we get in return – something psychologists call 'loss aversion'. This is part of the reason we hold on to things that we have no use for but don't want to be parted from.

You may be totally happy with your loss aversion – or you may feel it has turned you into a hoarder, and it results in a house full of things you never use or look at, but can't bring yourself to throw away. If you are comfortable with the situation, it's not for me to tell you to declutter, but if you would rather keep on top of the power that ownership has over you, there are a number of psychological tools you can use to help:

- **Two out, one in**. For some, getting too much stuff is driven by an enthusiasm, or even an addiction, to shopping. A useful technique is only to allow yourself the pleasure of buying an item if you get rid of two similarly priced items first. Leave the disposal until after the purchase and you won't do it – but with the incentive of the shopping coming afterwards it's much easier to part with things.

- **Add value to disposal.** One of the problems of getting rid of things is that you don't see a direct positive response to doing so. The previous suggestion was one incentive – another is to see an end result to your action that you will value. If you are a strong supporter of charity, don't just give things to the charity shop, but get a feel for what the charity will do with the money they raise from the items. Alternatively make use of a site like eBay to resell your items and envisage what you can do with that money. To give it more weight, add up your expected returns from getting rid of things over the next year – what could you do with that? Alternatively, use a freecycle scheme where someone else can make use of your unwanted item.

- **Make it easy.** It can be hard work to get rid of things. In some cases, loss aversion is driven as much by the faff involved in disposal as it is by the attraction of the item. So find a way to make disposal easy. Use the 'we'll take your junk away' type charity envelopes that come through the door. If you have a lot of clutter, get a skip. This might seem overkill, but there are two real incentives to having a skip. One is that it is trivially easy (and quite rewarding) to hurl things into it. The other is that to get your money's worth, you will want to fill it, encouraging you to be firmer with your disposal policy. A small skip from a local company doesn't cost much. This approach might offend the recycler inside you – but some problems require drastic surgery.

Salespeople use the power of ownership all the time. They offer you a money-back guarantee, because they know that once you own their product you will not be inclined to return it. Or they use ownership to get you to buy into things that involve spending money later down the line. Why does a satellite TV company or the internet company offer such amazingly good starting rates? Because once they've got you – and more to the point, you own their product – you will be reluctant to stop owning, and so will continue to pay the much higher later rates.

Generally speaking, we give in far too easily to the power of ownership. I have experienced a roadside car rescue service, insurance companies and a mobile phone company who, at the end of the initial period nearly doubled their fee or offered a poor renewal contract compared with their 'new customer' offers. In both cases, when I rang up and said I was leaving, they immediately offered the new customer benefits for another year. Often it's not that they are offering the service cheaply to get you in, but rather they push up the rates to an extortionate level once they've hooked you into ownership.

Challenge the price hike and suddenly it is their ownership of you as a customer that is in danger, and you get the upper hand.

LINKS:
- **Bargaining** – page 219
- **Habits** – page 226

P

• • • • • • • •

Price

As all good marketers know, price has a huge impact on psychology. If you are to fight back against the sales people, it's good to know what their tricks are.

When it comes to pricing, a good company has customers eating out of its hand. It uses a range of techniques to help part you from your money. Some are described in the **Advertising** section, but that's only the start of the onslaught.

We prefer things that cost more, even if they are no better quality than cheaper versions. You only have to look at the success of designer labels. Such a label is a way of telling other people that you spent more on your possessions than they are worth – yet many of us love them. But it's not just a matter of establishing status. If you tell someone a wine is expensive, they will think it's better than one you say is cheap – even if it's the same wine in both bottles. The taster will construct an experience to justify the price. Similarly, if you try out placebo painkillers – just sugar pills, but presented as painkillers – they will reduce the pain, because the brain assumes they are working. Give someone a placebo that costs £1, and it will have better painkilling properties than the same pill priced at 10p.

It's fine to pay extra for something you enjoy, whether it is the cut of a particular jacket, or the ambiance of an upmarket supermarket versus a cost-cutter. But be aware of what you are paying extra for, and what the seller is doing to you.

Another price issue is our tendency to round down. We are often our own worst enemies as shoppers. If we see an item priced £29.99 it feels a lot cheaper than one at £30.00 despite only being a penny less. Not only do we ignore the 99p, bringing the imagined price down to £29, we also tend to think in round tens, so our feel for the price will be more like the £20 mark. (And we will definitely say it cost about £20 to our partner.) Get into the habit of rounding prices up. When you see £29.99, think £30. This will give you a much better feel for the real price.

In basic items, cheap can be good, but *too* cheap is dubious. Most of us want to save money, but there comes a point where something is so cheap that we become suspicious. If, for instance, a sausage costs 10p, we have to ask ourselves, given the cost of production, packaging and transport to the shops, how much decent meat is in it – and the answer has to be 'not a lot'. Having a sense of what things cost to make can be very helpful in doing this. A 10p sausage should raise the eyebrows, but a 10p cup of tea is far more readily achievable – we do it at home every day.

LINKS:
- **Advertising** – page 212
- Placebo – see **Blind trials** – page 262
- Price and large numbers – see **Scale and understanding numbers** – page 239

Procrastination

**If there is one subject authors are
expert in, it is procrastination – but it is
something everyone suffers from.**

The late, great Douglas Adams, author of *The Hitchhiker's Guide to the Galaxy*, was a world-class procrastinator who said: 'I love deadlines. I like the whooshing sound they make as they fly by.' All authors are familiar with endless attempts not to start writing

– whether it's checking letters and emails or being distracted by the sudden need to look up a reference. Yet once they get started many of them enjoy the writing process. Only to find that after a break the procrastination starts all over again.

From a psychological viewpoint, procrastination is about weighting. The more immediate something is, the more weight we give to it. It's why politicians have such trouble with long-term projects, and why most of us don't give enough consideration to our pensions. What we are usually doing in procrastination is giving more weight to a small, short-term reward (I get to see what's happening on Facebook) than to a much larger, long-term reward (I'll have finished a book and get paid).

Of course, in some cases we are putting off an activity that is genuinely boring or distasteful. It doesn't take much psychology to explain why we put off cleaning the toilet and have a cup of tea instead. But it's a much more subtle balance when both activities are positive, but one delivers instantly, while the other delivers over the long term. If anything, current popular culture glorifies a form of procrastination, where we are told that the traditional approach of learning a musical skill (say) over a long period of time with many hours of practice can be supplanted by the immediate gratification of appearing on a reality TV show and becoming famous for the sake of being famous.

So how do we get around procrastination? We need to discover the reward mechanisms that make the short-term alternatives appealing, and better them. Don't look at a long project you keep putting off as the whole thing, because that is too distant. Instead break it down into bite-sized chunks, so you can reward yourself in a small way by achieving a goal each time you have a go at it. With book writing this is relatively easy, by setting a target for a day or week – and the same can hold with anything from decorating the house to starting your own business. You might think that it's the breaking into bite-sized chunks that's the important bit, but actually it's the satisfaction of getting something completed (bolstered by rewarding yourself) that makes the difference.

Another important step is to make the process visible. Not only do we give more weighting to immediate issues, we also do to the ones that clamour most for our attention. This is perfectly natural, but not always sensible. Think of our difference in feeling about the safety of air travel (see **Air versus road**). A major reason most of us are a lot more nervous about air travel than we are about travel by car is that there is a lot more media coverage of a plane crash than of a car crash.

What lesson can we learn? If you want to increase the weighting of something you are inclined to procrastinate on, make it more visible. Put it in your diary to do something about it. Leave visible reminders of it around the house. Keep in sight a piece of paper documenting the benefits you are going to get out of doing it. Have a very visible checklist of achieved milestones on the wall.

This sounds corny, like having a star chart for your children, but psychological tools don't stop working just because we have become adults. The more you can make a task and your progress visible, the less you are likely to put it off.

LINKS:
- **Air versus road** – page 214

R

. . . .

Risk

**Handling risk is an important part of dealing
with the world around us, which makes it
unfortunate that we are so bad at handling it.**

All life involves balancing risk – but unfortunately, risk is all about
probability and statistics, and human beings are particularly bad at
dealing with this kind of mathematics. For whatever reason it just
doesn't come naturally to us. Yet getting our response to risk right
is very important. Science can help us – but only if we are prepared
to take a step back and think, as our knee-jerk reactions on risk are
almost always wrong.

One aspect of the presentation of risk we particularly need to
watch is the use of percentages, as they can be highly misleading
if we aren't told actual numbers to put them into context. Say, for
instance, you were about to visit a city and you heard on the news
that it was experiencing a crime wave. Specifically, the number of
homicides in the city had increased by 100 per cent in the last year.
That sounds really bad – perhaps it might be better to put off your
visit. But you might feel quite differently if you heard that there was
one extra murder this year – the numbers had gone up 100 per cent
from one to two. This is a small number.

To put it into context, London has about 100 murders a year,
while New York manages 300–400 a year. But even those numbers
are misleadingly high, as very few of them involve tourists. The risk
is very low, but it is all too easy to get spooked by badly presented
statistics.

Research has shown that we give bigger weighting to losing
something that we already have, and one impact of this is that we

are more concerned about negative risks than positive benefits. Any action involves a mix of risks and benefits. So, for instance, children who play outside, climb trees and experience the world have been shown to be healthier and better developed than those who are kept indoors and not allowed to risk the inevitable physical pitfalls of being out in the world.

As we have seen in the **Air versus road** section, we also give artificially high weighting to issues that are given heavy media coverage. Because the outcome of child abduction, for instance, is so horrific, and gains so much media coverage, the extremely small risk of it happening to any particular child has resulted in extreme caution that prevents children from gaining the benefits of playing unsupervised and gaining experience that will help them in later life. It doesn't help that an increasing blame culture has made politicians and officials overstress risks. While media reports of 'health and safety gone mad' following a school banning schoolyard games, for example, are often overinflated, there is no doubt that we could all do with far better information to help us make sensible decisions in areas where risk is involved.

LINKS:
- **Air versus road** – page 214

Ritual

**Rituals have a bad name, but they are a
powerful aspect of human nature and can
be effective in lowering stress levels.**

Generally, we are down on rituals these days, as if somehow they were a remnant of a primitive past that we have advanced beyond.

If someone says 'This is getting to be a ritual with you,' they usually mean it negatively, as if you were stuck in a rut. And of course anyone who does everything inflexibly is not going to cope well in a fast-changing world. But to ignore rituals entirely is to miss out on a valuable stress-relieving tool.

The fact is that having a small number of rituals is a great way to cope with said bewildering, fast-changing world. Research on 'coping strategies' has shown that individuals tend to have higher self-esteem and better ability to cope with stressful situations when they have various kinds of ritual in their personal armoury. It's not that you should work through everything using a dusty old ritual, like the inhabitants of Gormenghast Castle in Mervyn Peake's books, but having a small core of ritual can provide a valuable defence.

If you are thinking 'I'm not religious,' this has nothing to do with religion. A ritual can be anything that is a simple comforting act that anchors your life. It could be enjoying a glass of wine with your feet up or reading the children's bedtime story. It could be watching your favourite soap opera. Of course, it could also be a religious service, and that is where the largest amount of research has been focused, but that's just one option.

One important thing in helping rituals to relieve stress rather than create it is that you must be prepared to be flexible and occasionally give up your rituals – but having at least one or two a week can make all the difference. What the activity is has to be down to you. It really doesn't matter, as long as it doesn't involve anything too strenuous and it is essentially repeated from occasion to occasion (though variations are fine).

On weekdays, rituals often work best in the evenings, as an opportunity to unwind after work, though at the weekend time tends to be more flexible and it might be sitting down with the Sunday papers for an hour.

Have a think through your week, consider your ritual armoury and see if it needs reinforcing.

LINKS:
- **Habits** – page 226

S

. . . .

Scale and understanding numbers

One of the ways we are often caught out by the limits of our thinking is in dealing with large scales. We didn't evolve to deal with big numbers, and so it can help if we put them into proportion and are wary when they bias our thinking.

In the everyday world we don't directly experience big numbers. Most of us have between zero and ten children. We rarely have more than ten real friends or 100 acquaintances (I'm talking about the real things, not Facebook friends). We experience the world one day, one meal at a time.

So when we are presented with very large numbers, the scale can throw us and mislead us. This often happens in politics, when a government throws around large numbers. What, for instance, does an extra billion pounds in taxation mean? It sounds an awful lot, even if it's for something we really believe in, like the health system. It can be helpful when suffering from this kind of scale problem to put the number into personal context.

There are around 30 million taxpayers in the UK, so if this were a flat per person tax, we would be talking about roughly £33 per person per year. Suddenly it doesn't seem quite as horrendous.

It's particularly important to be wary of large numbers when making comparisons. When buying a house, for instance, we might be casual about the difference between £200,000 and £205,000. But without the big numbers to distract us, £5,000 is a large sum of money, and one that you would be prepared to spend some time ensuring you got.

An interesting US study asked people how much they would

go out of their way to save money. Most said that they would take a fifteen-minute walk to another store to buy a pen at $18 rather than $25. Yet most would not waste that time to get a suit at $448 instead of $455. The difference in that case didn't seem worth the effort. But the saving – $7 – was the same in each case. The decision really should have been based on the answer to 'Am I prepared to walk 15 minutes to save $7?', but the scale provided by the high cost of the suit hid the real nature of the decision.

Whenever large numbers are thrown about, make sure you think about savings or extra costs in terms of their real impact, rather than being dazzled by the scale of the numbers. Think about what you do with the money that could be saved (say) in other ways to put it into some sort of context. It can also be useful to think of it as an hourly rate. So in the American example, it would be useful to think 'Is $28 a reasonable payment for an hour of my time?' – because the person taking the walk is effectively being paid $7 for 15 minutes of 'work' – and to act accordingly.

Another scaling trick works in the opposite direction. If you are expected to pay for something regularly and you are told it only costs, say £2 a day, make sure you put this into the timescales over which you usually consider your income – typically monthly and annually. While £2 a day doesn't sound much, £60 a month sounds more substantial, and £720 a year could be used for all manner of interesting things. Make sure you adjust the scale to work with more familiar kinds of number.

Self-esteem

**One of the most significant factors in
success or failure – and in reducing stress
– is self-esteem. But with awareness of
this, it's simple enough to push it up.**

In a famous piece of research, British civil servants at different levels
in their organisation were compared. It was found that those in the
'high stress' important jobs had significantly *lower* levels of stress
than those doing the menial jobs. The two factors that seemed most
to influence this surprising reversal were that those in the top jobs
had much more choice over what they did, and that they had higher
levels of self-esteem. What's more, that self-esteem seems to make a
significant contribution to an individual's success in life.

One of the problems with low self-esteem – which afflicts a
good proportion of the population – is that it is self-reinforcing.
If you think that you never succeed, then you are less likely to try,
and you feel bad about your failure, increasing stress and resulting
in even less achievement. Research has shown that we acquire self-
confidence and self-esteem as a result of successful experiences, but
often, as adults particularly, we tend to ignore the successes when
feeling low and focus on the negatives. Giving 'failure feedback'
reduces future success and self-esteem, while giving 'success feedback' has the opposite effect.

Three quick exercises over a week could help transform your
opinion of yourself.

On day one, at the end of the day, note down three to six successes you have had that day. It doesn't matter how small they are.
It might just be 'Got out of bed on time,' 'Saved £5 on the shopping
with special offers,' 'Finished something at work,' and 'Read the children a bedtime story.' Repeat this for a total of five days.

It doesn't matter how bad a day it is, you can always find some achievements. Don't try to compare them with what went wrong – simply think of the positives. The point of this part of the exercise is to overcome the feeling that everything goes wrong all the time. It really doesn't.

On day six, take a little longer to think through your life to date and list a few of your biggest successes. These can be things that are important just to you – perhaps family landmarks, exam success, getting a job, learning to drive – or things that have a wider impact, achievements that other people might consider special. When you have the top handful for your life so far, think back to when they happened. Remember what you felt like at the time. Enjoy those moments again. Some of us feel guilty about enjoying success. We feel that it is somehow vulgar to do this. But that is a ridiculous attitude. You deserve that warm feeling. And bringing back those good moments will have a positive effect on your current self-esteem. You've done it before and you can do it again. You are a winner.

This may sound artificial and forced, but psychologically we are programmed to respond well to praise and reward, even if we deliver it ourselves. Sometimes that cycle can go wrong – it's essentially what happens with comfort eating – but here it is being used in a positive way to lift your self-esteem.

On the final day, look forward. Look for opportunities to succeed. Plan a little success. It won't all work. Most of it may not. But taking a positive outlook for the future will also help that self-esteem. Don't think, by the way, that this is the same as the cult of the individual that emerged from the 1960s and resulted in people being extremely self-centred as they trampled over others to 'find themselves'. The stress relief from building self-esteem will help your friends and family just as much as it does you. If you have low self-esteem and are stressed you will have a negative effect on them too.

Smiles and other body language

Many of us believe we are experts in reading other people, and there are some unspoken communications that are hard to fake – but a smile isn't one of them.

'Her smile was so fake!' If someone demands a smile of you, it can look very unnatural, but experiments have shown that we are very good at faking a smile outside of unnatural conditions like posing for a photograph.

Some body language – the unspoken communication that is going on between us all the time – is hard to simulate. But the smile is a relatively easy one. In tests performed at the University of California, a whole group of experts from psychologists to secret service agents and customs officers tried to determine which of a group of nurses were really smiling and which were faking it as they described something they had seen. Pretty well all the experts were fooled.

Some other basics of body language can also be learned. So, for instance, if you don't want to look nervous in an interview, maintain eye contact and sit with relaxed, open arms, rather than folding them. Lean slightly forward to show interest. This is all easy enough to fake. But what we can't do is fake all the tiny muscle movements that accompany our thoughts – including, for instance, the way the pupils of our eyes dilate when looking at something we find attractive. So it can still be difficult to fool an expert in other ways.

Standing out

Although we humans like to go around in packs, we also feel a need to stand out from the crowd, a conflict that crops up everywhere from a restaurant to social media.

There's nothing better than going out with friends for a meal, but even though you may all get on well, you also have a natural tendency to want to stand out from the crowd and not to seem to be following the herd. Studies have shown that when people make a choice from a menu, whether it's for food or drink, and hear what other people have chosen first, they are much more likely to go for something different to the others. This even extends to ordering something they don't really want – or certainly don't want as much as a popular choice – if it prevents them from looking like a sheepish follower.

It might seem that this is because the group around the table are going to share to try out each other's choices, so want a more varied selection – but outside of particular cuisines where this is the norm, there is no evidence of this happening; people just grimly eat their substandard choice.

There is a simple way to avoid this. When selecting from a menu with a group, make sure you choose what you want before any discussion of what people are going to order – and once you make your choice, stick with it. Avoid the temptation to switch away from what you really wanted in order to maintain a difference and you'll have a more enjoyable meal. Sometimes psychology means a degree of tricking yourself to get what's best.

In recent years we have had a whole new opportunity to study human attempts to stand out from the crowd in the way we use social media. There is a clear distinction between uses of Facebook, Twitter and the like to keep up with friends – the reason these services were set up – and for self-promotion. There have been some studies of the way that celebrities use Twitter in particular to share personal information to reinforce their celebrity status. This practice seems to have encouraged others, who don't have a natural group of followers, to become 'trolls', aggressively attacking others on social media to make themselves stand out, generating a form of artificial celebrity. As yet there has been relatively little work done on the psychology of those who misuse social media, but it seems that, as in the real world, the best way to stand out in social media is not to create artificial celebrity but rather to achieve something of value in its own right that will bring with it personal distinction.

Swearing

Most of us were brought up to think of swearing as in some sense 'bad' – yet most of us do it sometimes, and in the younger generation it has become more acceptable. But does it have a benefit?

Older folk might still get rather upset by swearing, but it does seem to perform a useful psychological function – at least in some circumstances. Research on men carried out in 2009 suggested that there is a good reason that we tend to yell expletives when we hurt ourselves. By comparing the effects of swearing against using everyday words, it was discovered that yelling swear words increased the ability to tolerate pain and decreased the amount of pain that was felt.

This effect wasn't felt by men who tend to be overly-dramatic as a matter of course, reflecting, perhaps, the observation that swearing loses its impact when used most frequently, and that someone who is generally thought not to swear can really shock by dropping in a swear word. The suggestion from the research was that swearing could break the link between fear of pain and the feeling of pain, reducing self-induced suffering.

U

Upselling

If you've never worked in a sales job, you might not be conscious of upselling, but you have certainly been on the receiving end of this psychological technique.

In the early days of American fast food restaurants in the UK, you may have heard the question: 'Would you like fries with that?' Today it would seem a little odd, because we are so familiar with the product, but initially this appeared a useful reminder that a burger didn't come with fries unless you ordered a meal. However, from McDonald's viewpoint, it was an opportunity to make some more money, because upselling is about piling stuff on once your customer has already made an initial commitment.

Here are some more examples of upselling:

- You buy a new TV and you are asked if you would like an extended warranty.

- You buy a new car and you are asked if you would like floor mats, special paint treatment or a prepaid servicing contract.

- You travel on a budget airline and you are asked if you would like to pay a little extra to insure the cost of your ticket against cancellation, choose your seat or put luggage in the hold.

Even 'buy two, get one free' is upselling, because otherwise you might only have bought one.

Upselling is good from the seller's viewpoint – and even from your viewpoint it might be a good thing, but you have to be wary, because the seller has you at a disadvantage. Particularly when you are paying for a big-ticket item, like a TV set or a car, the psychological impact of scale means that you don't think of the additional cost of the warranty or paint treatment as significant.

To see if the upsell is doing you a favour or taking you for a ride, mentally detach the extra from the main purchase. Imagine that main item had already been bought. Now think of the extra in its own right. Is it worth what you are being asked to pay for it, and what are the chances you will benefit from it? Extended warranties are particularly interesting in this respect. Often they are a waste of money. Something like a TV that has no moving parts is likely either to break down in the first year, while under the manufacturer's warranty, or to last longer than the extended warranty period.

Where an extended warranty is worth considering is an expensive purchase with mechanical functions that can go wrong, like a high-end fridge or washing machine, or a piece of technology that is carried around, will be damaged when dropped, and that you would struggle to be without, like a phone or a laptop.

If we look at the typical upselling with a car, mats are often remarkably overpriced – yet we tend to throw them in because £60 or £100 for four bits of fabric is neither here nor there when spending thousands on a car. But it's still worth thinking about whether you need them at that price – and the same goes for the paint treatment. Prepaid service contracts for a car are more likely to be win-win and worth considering. The dealer likes them because they tie you in to having your car serviced there. But if you were going to do that anyway, you benefit because you can get the servicing significantly cheaper than you otherwise would.

The trick, then, when someone is trying to upsell to you, is to detach the extra from the original purchase and think of it as a separate item. That way, you can separate the upsold sheep from the goats.

LINKS:
- **Scale and understanding numbers** – page 239

Using incentives

We regularly use (and are the targets of) two types of motivation – social and monetary – but the two don't neatly work together.

Whenever we try to get someone else to do something – or, for that matter, someone tries to persuade us to do something – we use incentives. Broadly, these divide into social and monetary incentives (in which I'm including other market transactions as well as pure cash). We can get in a real fix if we bring the wrong type into play.

If we engage a monetary response in a social situation, it causes offence. If, for example, you visit friends for a meal or go on a first date and say at the end of it 'I had a great time, here's £50 for your trouble,' things will not go well. Instead, in a social setting we expect a social incentive – thanks, praise, offers to help with the washing up or whatever, rather than cold, hard cash.

Gifts are a strange, in-between type of incentive. Experiments have shown that as long as money isn't mentioned, and the gift isn't too extravagant, it is treated as a social incentive. But if it is ridiculously out of proportion (giving someone a car on a first date, say) or if money is explicitly mentioned ('Thanks for that lovely meal, I'm giving you a £5 pot plant as a thank-you'), it turns into a monetary incentive and sends the wrong message.

In broad terms, social incentives provide better motivation than monetary ones if the stakes are not too high. Studies have shown that people will perform better when asked to do a task for a favour than when offered a relatively small amount of cash. But there is a clear switch, as can be seen with a psychology experiment known as the ultimatum game.

Two people take part and are offered some cash – a pound say – to share between them. Without any communication, the first person decides how to split the money. He or she can do it any way – all to him- or herself, all to the other person or split any way in between. Then the second person says 'Yes' or 'No'. If it's 'Yes' the cash is split the suggested way; if it's 'No' neither of them gets any money.

Logic and the monetary incentive suggest that the second person should say 'Yes' even if they are only offered one penny from the pound. After all, it's money for nothing. But in practice, they will usually say 'No' to anything under around 30p – a clear example of social incentives being stronger than monetary.

However, there is a twist in this game. Things would be very different if the total stake were £10 million. In such a game, if you turned down a 1 per cent offer, you would be refusing £100,000, which is sufficiently life-changing for most of us not to refuse. I use this as an exercise in talks, asking the audience all to stand up, then to sit down when I reach a figure they *would* refuse. Interestingly, some sit down as the gradually reducing offer gets to around £5,000. But most hang on until we're down to something between £200 and £50 – a tiny fraction of the total cash. Social incentives win to a point, but most of us are prepared to go for the money if there is enough on offer.

Since a company can't spend large sums of money on you to get you to buy their products, you will often find that they mimic social incentives in the way they advertise themselves or in the way they provide incentive schemes. Companies will, for instance, try to make you feel part of their 'club' or run advertising campaigns where customers choose a new flavour of product or are made to feel special by being part of a select grouping. They are also making increasing use of social media (the clue's in the name) to give the impression that we have a social connection with them – much more powerful than simply offering a penny here and a penny there.

This seems to be why we get so upset when companies respond mechanically and financially when things go wrong. We get really irritated, for instance, if a phone company slaps on a bill if we go a little over our prepaid tariff, or the bank charges us a hefty fee for dipping briefly into an overdraft. But that's the price they pay for attempting to give social incentives. You can't mix social and monetary motivations without causing pain.

LINKS:
* **Scale and understanding numbers** – page 239

HEALTH

• •

Diet is one major area where science can give us guidance on health concerns, but there is also the whole range of treatments we undertake, from complementary medicine through to modern pharmacology.

Having an understanding of whether an alternative therapy, say, has a scientific basis or is pure mumbo jumbo can be a useful factor in deciding whether or not to buy it – and we will provide that information. But of course it is also possible that a therapy can have benefits that aren't understood by science yet. For instance, the whole idea of vaccination – apparently giving yourself a mild dose of an illness, or another illness that is related to it – originally seemed crazy. Now, though, we know that there is a good scientific basis, because the vaccine stimulates the body's own defence mechanisms to be able to fight off an invading virus or microorganism.

Luckily, even when there isn't any science that we can use to establish *why* a treatment might work, science can still help us out with whether or not a treatment really does have any value. All too often the benefits of a treatment are anecdotal. We've heard that, say, a friend felt better after taking a homeopathic treatment, so clearly it works. But there are many reasons someone might feel better, and most of them aren't down to taking a pill.

Science gives us a process called the double blind trial that makes it possible to determine whether a treatment really does have a benefit, even if we don't know how that treatment works. See the entry on the **Placebo effect** (page 315) for more information.

A

. . . .

Acupuncture

**Acupuncture is an alternative medical practice that
currently has some National Health Service funding
in the UK. But is there any scientific basis for it?**

When looking at any alternative medical treatment, there are three
questions to be answered. Does the explanation of the therapy
make any sense? Is there good scientific evidence for it working?
(Which could be the case even if the explanation doesn't make
sense.) Is it a placebo and, if so, is it acceptable to use it for the
placebo benefits?

The original basis of acupuncture is that a person's health is
dependent on a life force known as *ch'i*, which flows through the
body along routes known as meridians. The reasoning is that meridians can get blocked, and acupuncturists place fine needles in the
patient at the critical points, clearing blockages and ensuring free
flow of the energy. This has zero scientific basis. There is no evidence for the existence of *ch'i*, there is no evidence for the existence
of meridians, nor is there evidence that the placement of needles
can in some way free up the flow of the non-existent energy in the
non-existent meridians.

However, herbalists first used willow bark (which contains a
painkiller related to aspirin) medicinally with an incorrect explanation. It is possible for the explanation of the workings of an alternative therapy to be rubbish, even while the effect is actual and
beneficial; so we need to look for verification of effectiveness too. It
is worth bearing in mind, though, that unlike simply taking a herbal
medicine, the basis for acupuncture does depend on the layout of
a non-existent network of meridians.

It is generally agreed that acupuncture originated in China, dating back over 2,000 years. It started to be introduced in Europe in the mid-19th century, but faded out in the West until Mao Tse-tung encouraged the reintroduction of traditional medicine in China in the 1950s, and it passed from there to America in the 1970s. Support from a World Health Organisation (WHO) document in the late 1970s, which claimed acupuncture was effective on everything from the common cold to duodenal ulcers, fuelled a major growth in the West.

Early tests of the value of acupuncture were worthless, as they did not use double blind trials. This meant that the results could simply be reflecting expectations of the patient – the placebo effect. However, the 21st century saw a range of blinded trials used. It is harder to perform 'pretend' acupuncture than it is to use a sugar pill in blind trials of medication, but the best trials reported in the Cochrane review (described below) were done comparing real needles with fake needles that didn't make the crucial penetration, and with real needles that were placed at points other than those considered essential by acupuncturists.

A report issued by the WHO in 2003 seemed to confirm their early support, but the majority of the trials it relied on were not of good enough quality and adequately controlled. Many of them have since been demonstrated to suffer from publication bias, a problem where good results are cherry picked and negative results ignored. This isn't surprising when it is realised that the entire panel assessing acupuncture for the WHO were 'true believers' – there was no one who was sceptical about the technique.

A more recent Cochrane review – the most respected assessment of reliable medical evidence, taking only good quality trials into account – made it clear that there is no evidence as yet of acupuncture being effective over the benefits of a placebo in the vast majority of problems it was claimed to be effective for. There were a small number of cases, notably lower back pain, where the review concluded that the value of acupuncture was supported, but this was with the proviso that the quality of the evidence was not good enough in these cases. More recent high quality studies, using a better, more convincing fake needle, have suggested that even these remaining areas of doubt were the result of the placebo effect.

Despite the UK National Health Service currently still paying for some acupuncture treatment, it seems to be nothing more than a placebo.

LINKS:
- **Placebo effect** – page 315
- Double blind trials – see **Blind trials** – page 262

Allergies

Allergies are examples of the body's defence systems making an error, which at worst can be extremely dangerous, but more often just causes irritation.

Your body has a complex set of mechanisms, collectively known as the immune system, to protect you from attack by viruses, bacteria, fungi and other assailants that could otherwise make you ill or even kill you. The immune system is not a single process, but a whole armoury of different defences. There are static defences like the skin and the blood/brain barrier – the physical mechanism that allows water and essential chemicals through to the brain but prevents potentially dangerous molecules getting through. And there are active defences, like the wide range of white blood cells that take the attack directly to would-be dangers and engulf them.

This complex system generally works well, but it can get things wrong, relying on specific triggers that can be set off by innocent factors. In an allergy, something you breathe or eat is assumed to be an attacker by the immune system, producing anything from basic symptoms – a runny nose, sneezing, red eyes and itchiness – to life-threatening reactions like the anaphylaxis sometimes triggered by insect stings and food allergies, where white blood cells overreact, producing swelling that can be uncomfortable or even cut off the airway. In between these extremes are allergic responses that can be painful and disfiguring but rarely life-threatening, like eczema.

Science has given us a number of responses to allergic reactions. Anaphylaxis is usually treated using adrenaline, the hormone the body makes use of to kick systems into action for fight or flight. Those who live or work with people at risk of serious allergic reaction will be familiar with the EpiPen, an automatic injection device to provide an adrenaline boost (named after the American name for adrenaline, epinephrine). Mild allergies like hay fever are likely to be treated with antihistamines, drugs that block the action of histamine, a chemical in the body that is central to the allergic response.

The numbers of people affected by allergies have been increasing by about 5 per cent year on year, particularly among children. The charity Allergy UK suggests that around one in four people in the UK will have an allergy at some point in their lives. (Allergies can develop later in life or fade away with time.)

There are a number of factors that influence our chances of developing an allergy. Some appear to be genetic – if your parents have some allergies, you are more likely to as well. Other aspects are environmental – if you are a child in a house with above average levels of the typical causes of allergies, like cigarette smoke, dust mites and pets, it increases this risk, as does too much use of antibiotics.

There is not a good scientific consensus on why there is such a significant increase in allergy sufferers. Some have put this down to pollution, though allergy rates are growing in countries that have much lower pollution now than they did 100 years ago. Others believe that we bring up children in too sterile an environment, giving the immune system 'too little to do' in dealing with bacteria, so it becomes over-sensitised – but this seems an over-simplistic picture. Whatever the reason, the increase in prevalence is real.

It is not uncommon for allergies and food intolerance to be confused. Allergy is an unwanted stimulation of the immune system by a substance that could be a food (peanuts, for instance). A food intolerance like lactose intolerance or gluten intolerance is where the gastrointestinal system has a problem with digesting a particular kind of food, which can result in bloating, nausea and some allergy-like symptoms. But an intolerance is not an allergic reaction, and someone who is, for instance, lactose intolerant is not 'allergic to dairy products'.

LINKS:

- **Immune system** – page 301

Antacids

**When we get a stomach ache – acid indigestion,
as the advertisers like to call it – we reach for
the antacids and indulge in basic chemistry.**

Your stomach has some serious acid on board – hydrochloric acid. It's an acid that is strong enough to inflict serious damage. But that is just what is needed in your stomach. The job of that acid is to break down whatever you eat so that it can be used to generate energy, as well as making it easier to dispose of the waste.

The acid levels in the stomach vary and if they are too high acid can attack the stomach wall, resulting in discomfort. So can the action of the stomach acid on various parts of your insides it shouldn't reach, for example when there is reflux, where the acid sprays out of the stomach and up into the oesophagus producing 'heartburn'. These problems are most commonly a result of bad eating habits (such as overeating or eating too late at night), although some people suffer from them as a result of physical problems like a hiatus hernia.

The instant remedy? Pop in an antacid tablet. The result is a simple chemical reaction. Although there are various kinds of antacid on the market, many contain a carbonate like calcium carbonate or magnesium carbonate. Calcium carbonate is a very common mineral. It gives eggshells their solidity and it is the main constituent of limestone, marble and chalk. You are, in effect, eating a powdered rock when you take an indigestion tablet, though I wouldn't recommend this as a cheap substitute. When an alkaline chemical like calcium carbonate meets hydrochloric acid, there is a chemical reaction, acid levels drop and your stomach becomes less uncomfortable.

More powerful prescription drugs have the impressive name 'proton pump inhibitors'. They are the most commonly prescribed medication and act by suppressing the mechanism that increases acid levels in the stomach. Rather than cancel out the existing acid, like a traditional antacid, they prevent more acid being produced.

Antibiotics

Arguably, antibiotics have saved more lives than any other treatment, killing the bacteria that cause deadly diseases. But we should be careful that they are only used when essential.

When antibiotics were first introduced they were miracle drugs. Targeted to kill the bacteria that cause many diseases, they transformed medicine and made long, complex operations possible without significant risk of infection. This meant that early on doctors tended to prescribe antibiotics for pretty well anything – and patients, who perhaps had seen the impressive way that amoxicillin finished off an ear infection, began to demand antibiotics every time they had as much as a sniffle.

There are two problems with this. One is that antibiotics only work on bacteria – and many diseases are not caused by bacteria. Colds and flu, for example, and many common causes of sore throats (as well as a range of serious diseases) are caused by viruses, on which antibiotics have no effect. Yet there was a period when doctors prescribed antibiotics anyway, at their patients' request.

Secondly, and more seriously, bacteria are the fastest evolving creatures around, in part because they reproduce very quickly and in part because they can pass DNA between each other. Should one bacterium, by random mutation, develop a resistance to an antibiotic so the antibiotic doesn't kill it, this resistance can spread very quickly, not only within a particular species, but jumping from one set of disease bacteria to another.

This ability to resist has been growing faster than we can come up with new generations of antibiotics. Developing these drugs is a long, slow process, and doesn't pay as well initially as developing, for instance, anti-cancer drugs. As a result, we are nearly at our limits, with our antibiotics of last resort in danger of failing. This is the problem with MRSA, the 'superbug' that is resistant to the powerful antibiotic methicillin.

There are potential alternatives to antibiotics, notably 'phages', which are bacteria-killing viruses, which have never received the same level of research as antibiotics. But for the moment it is essential that we keep use of antibiotics down to a minimum. This is particularly a problem in agriculture in some countries, notably the USA, where antibiotics are given to animals routinely, both to avoid infection and because they promote rapid growth. This 'giving out antibiotics like sweets' is a major reason for the fast development of antibiotic-resistant strains of bacteria and should be a concern for us all.

Aromatherapy

It would be difficult to argue that aromatherapy is not a pleasant experience – but can it do any more than make us feel relaxed and soothed?

The concept of aromatherapy is to use fragrant essential oils (oils containing plant extracts, also known as 'essences', not medically *essential*) either as part of a massage or simply to give sensory stimulation to the person being treated. It is often used for aches and pains.

Although aromatherapists suggest that specific essential oils are effective in treating specific conditions, all the evidence is that we simply feel better after a touch of pampering with pleasant smelling oils, and the actual variety used has no influence.

There is no doubt that aromatherapy is a pleasant way to de-stress – but don't take it too seriously. Note, by the way, that Bach flower remedies, named after an early 20th-century homeopath, have more in common with homeopathy than they do aromatherapy, as the substances are so dilute that there are no significant active ingredients left. They are not the same as the essential oils used in aromatherapy.

LINKS:
- **Placebo effect** – page 315

B

. . . .

Bacteria

**Bacteria are the cause of many diseases – but they
are also important helpers for our digestive system.
One thing's for sure: they aren't going away.**

Bacteria have been on this planet for billions of years longer
than us, and they outnumber everything else to such a degree
that based on numbers of life forms alone, an outside observer
would see the Earth as a planet inhabited by bacteria with a few
other creatures.

Your body is made up of small packages called cells – around
10 trillion of them. Yet your body contains around 100 trillion bacteria. Many of these are 'friendly', in the sense they help the body, or
neutral – and even a lot of the decidedly unfriendly ones are not in
a position to do any damage. But occasionally bacteria will attack
and cause an illness. Some products attempt to add more friendly
bacteria in 'probiotics', but generally speaking these bacteria don't
stay in the system.

Advertisers of cleaning products encourage us to take a panic
approach to bacteria, which, they imply, are festering on every
surface in your house, waiting for the chance to attack. They
certainly are there in dangerous numbers when contaminants
get into the house, which is why we are encouraged not to wash
raw chicken any more, to avoid spreading them. And they will
spread if we are not careful, even with a quick contact (so ignore
the five-second rule). But, realistically, we are never going to
live in bacteria-free conditions, so need to concentrate on those
areas where they are most likely to get into our system and cause
us harm.

We should always wash our hands, for instance, after using the toilet and before dealing with food. And fruit and vegetables (especially if to be consumed raw) should be washed, while surfaces and chopping boards should be kept clean.

LINKS:

- **Antibiotics** – page 258
- **Chicken washing** – page 271
- **Five-second rule** – page 287
- **Hand washing** – page 289
- **Probiotics** – page 80
- **Vegetable washing** – page 324
- **Viruses** – page 325

Blind trials

The only effective medical trials are known as 'double blind'. These are essential in order to be clear if a treatment has a benefit.

When it is said that an alternative treatment like homeopathy is useless, many people argue: 'But I have taken a homeopathic treatment, and it worked for me.' The reason such anecdotal evidence can't be used to see if something really does work is twofold. One is that the evidence is often indirect. It's often not 'I have taken' but 'I know someone who took'. There is a good reason why hearsay evidence is not accepted in court. It is all too easy for information to get corrupted when it is an indirect account.

The other problem with anecdotal evidence is the placebo effect. If we compare the effect of, say, a homeopathic treatment and a sugar pill and know which is which, it is all to easy to say that the treatment is working and the sugar pill isn't. One reason for this is because we fool ourselves. If you ask ordinary wine drinkers which is better, wine from an expensive-looking bottle and wine from a cheap-looking bottle, they will say the expensive-looking wine tastes better, even if the two wines are identical.

But there is also a real side to the placebo effect. If we take something, for instance, that we think is an effective painkiller, our bodies release natural painkillers and opiate-like chemicals that reduce pain and help us feel better in ourselves. So just because a treatment makes the user feel better doesn't mean the treatment is actually doing anything.

To establish whether or not a medicine produces a real effect requires a blind trial. Here we conceal from the trial subjects whether they are getting a real treatment or a fake. If there is no clear distinction between the two across a large number of subjects then it is likely that the treatment is worthless.

However, there is one more problem. Surprisingly, if the people administering the medicine (or the wine) know which is the 'real' treatment and which is the fake, they unconsciously signal this to the trial subject. While the effect isn't as strong, even when the trial subjects are 'blind', some will be directed to detect the 'real' treatment. For this reason, a proper controlled trial needs to be *double* blind – both the trial subjects and the people administering the trial need to be unaware which is the medication under test and which the sugar pill. Only in those circumstances can we have good evidence that there is a real effect.

Double blinding should, in theory, be used for all scientific experiments. For example, physicists, trying to discover what is happening with fundamental particles, have been discovered to unconsciously bias the results towards the outcomes they expected if the data wasn't worked on blind. The only trouble is that while double blinding is easy to apply when the treatment involves pills, it is much harder to do with dietary changes and many other things that can influence our health.

LINKS:

- **Placebo effect** – page 315

BMI

Body mass index is a commonly used indicator of whether or not we have a healthy weight. It is a useful broad guideline, but has to be used with care.

The body mass index (BMI) is a simple measure based on your height and weight that is used to give guidance on whether or not you are overweight, underweight or a healthy weight.

An adult with a BMI below 18.5 may be underweight and should check with a GP. Between 18.5 and 24.9 is generally considered the healthy range. Over 25 you are considered overweight and over 30 becomes obese. (The approach to BMI used for children is more complex, based on how they compare with the population of the same age – see the calculator details below.) Being seriously overweight is a real problem. If you have a BMI of over 40, for instance, the risk of dying early shoots up, as there are significant increases in diabetes, heart disease and some cancers.

You can work out your BMI as an adult easily enough by dividing your weight in kilograms by your height in metres, then dividing the result by your height again. If that sounds too tedious, or if you would like to check the BMI of a child, there is an easy-to-use BMI calculator at this webpage: http://www.nhs.uk/livewell/loseweight/pages/bodymassindex.aspx.

However, should you discover a slightly raised BMI, this doesn't mean you should panic, especially if you put on a few pounds over Christmas. A major review, covering a vast number of participants, suggests that being slightly overweight according to the formal definition is better for you than being 'normal' weight. Traditionally, the aim is to get your BMI between 18.5 and 25, but the review showed that there was a small but significant advantage (a 6 per cent lower death risk) if your BMI was between 25 and 29.

The studies don't show a cause for this, though there is speculation that it could be because carrying a little fat helps hold off infection. It is also possible, though, that this is a misleading statistical effect. After all, those who do a lot of muscle building exercise will tend to have a high BMI (muscle weighs more than fat), and it is possible that some of those considered a normal weight have actually been heavier but dropped in weight due to illness. This is not a result that recommends being in that 25–29 range, but rather it means if you are in that range, it is less worrying than it sounds.

BMI is used because it is easy to calculate, but as a measure of healthy body mass it isn't particularly good. Many athletes are technically obese on their BMI alone due to their weight of muscle. The measure also goes wrong with people who are significantly above or below average height. It is entirely possible to be fit and healthy and yet be labelled obese on BMI. There are better measures, such as aerobic fitness, but they require much lengthier technical checking.

This limitation of BMI does not mean that you can ignore a high BMI. And most of us who are overweight know it. But should you have particularly well-trained and developed muscles, or be extra tall, it might be worth asking for a more comprehensive test.

LINKS:

- **Obesity** – page 307

Breast milk

There is good medical evidence that consuming breast milk has health benefits for babies. Although not always possible or advisable, where there is no reason to avoid it, breastfeeding is best.

There is now good scientific evidence to support the benefits to babies from breastfeeding. Though modern formula milks are very good, they lack some of the specific content of human milk. These include antibodies, enzymes and hormones which act to protect the infant.

Among the quantified benefits are reductions in chest and ear infections, reduced risk of vomiting, constipation and diarrhoea and reduced risk of developing eczema. There is also a proven reduction in the chances of becoming obese in later life.

Benefits also extend to the mother, who will have a reduced risk of getting breast and ovarian cancer. The process also uses up to 500 kcal (calories) a day, so helps the mother keep fit, reduces the cost of having a baby and requires less work to ensure correct milk temperatures, etc.

Some mothers have clinical reasons for avoiding breastfeeding – always check with your medical team.

LINKS:
- **Brain food** – page 158

C

· · · · · · · ·

Cancer

**We see more advertising by charities and more
news stories related to cancer than any other
health issue. Yet often what we hear is misleading.**

Cancer is a collective term for a health problem where cells in the
body begin to divide and grow out of control, forming malignant
tumours with a potentially deadly outcome. We hear a huge amount
about cancer, but some of it is based on myth. Sadly, we also often
see supposed 'cures' for cancer provided by everything from alternative medicine to vitamin C. While modern medicine is offering
more and more ways to combat cancer, these alternative approaches
can be dangerous if a sufferer avoids treatment in the belief that the
alternative approach can be successful. In some countries, including the UK, making claims to cure cancer without a sound medical
basis is illegal.

Some of those peddling alternative 'cures' come up with a whole
list of incorrect statements. You may read, for instance, that cancer is
a modern affliction, a result of our unnatural modern way of living,
and so we ought to return to nature to avoid it. Oddly, this is true,
but not in the way that it is meant. Cancer has existed throughout
history; it simply hasn't always been identified as such. But there
is no doubt that cancer is significantly more common now. This is
because one of the main risk factors for cancer is age. The older you
are, the more likely you are to suffer from it. And because far more
people live into their seventies and eighties than our ancestors did,
there are many more cases.

A lot of the misleading information about cancer is about the
impact of things we eat. As you will discover in the **Antioxidants**

and **Superfruits** sections, so-called 'superfoods' from blueberries to green tea are supposed to help protect us from cancer because of their antioxidant content. There is no actual medical evidence that this helps. This doesn't mean that these foods are bad for you – many are excellent contributors to a balanced diet – but they aren't going to make for a miracle cure either. Similarly, some quacks recommend avoiding 'acid' foods, as these are supposed to encourage cancers. There is no evidence of this whatsoever. The best things to avoid are alcohol and tobacco, which both have strong proven links to cancer.

The good news is that the majority of people won't get cancer at all – it's about 1 in 3 who contract it at some point in their lives – though as we live longer, this proportion is likely to increase. Another hopeful point is that the vast amount of money and effort going into cancer prevention and cures is paying off. Some cancers are now preventable, as for example with the use of the HPV (human papilloma virus) vaccine in teenage girls, which is proving highly effective at reducing the risk of contracting cervical cancer. Not all cancers are caused by viruses, but increasingly causes are being found that can be countered. And the survivability of many cancers has increased hugely thanks to modern treatment. We have a long way to go still, but we do seem to be making significant progress in the war against cancer.

LINKS:
- **Antioxidants** – page 7
- **Superfruits** – page 104

Carbon monoxide

**Smoke alarms have become the norm
in many houses, but carbon monoxide
detectors lag well behind.**

For some time now it has been no surprise to find smoke alarms in a house – and all new buildings in the UK have to have them. But carbon monoxide detectors are still relatively rare – yet they too play an important role.

When anything containing carbon (from wood to natural gas) burns it produces carbon oxides. The more familiar carbon dioxide may be the scourge of global warming, but it is not strictly poisonous. We can't breathe if there is too much of it in the atmosphere, but because it sinks to the ground and isn't produced in too large a quantity, it isn't a threat to our health.

However, carbon monoxide, which is produced when a fuel isn't burning efficiently, is actively deadly. It latches on to the haemoglobin that carries oxygen through our bloodstream more easily than oxygen does, so blocks our oxygen intake.

Carbon monoxide is most often produced in the home by faulty heaters – any gas fire, heater or boiler should have an annual checkup by a professional. But just in case, it is well worth getting a carbon monoxide detector, as the gas is colourless and odourless. There is no other way of being sure it is present, though if everyone in the household is suffering from persistent headaches, it is well worth checking.

Another potential source of risk is using outdoor cooking devices, particularly charcoal barbecues, which produce a lot of carbon monoxide, in a confined space.

LINKS:
- **Carbon dioxide emissions** – page 333

Catching a cold

We all tend to get colds at least once or twice a year and, despite living in a time of modern medicine, many of us still think of colds in a way that dates back before science and medicine had any close relationship.

One of the reasons there are so many useless 'alternative remedies' is that there was a time when 'real medicine' was so dangerous that doing nothing was often better than seeing a doctor. It's easy to forget that it was only in the 20th century that medicine moved from folk-tales to medicine with a scientific basis. Until then, for instance, doctors would routinely bleed patients – a procedure that made them weaker and more susceptible to illness – because they thought this would 'balance the humours', a concept based on non-existent substances in the body.

Back then, a lot of diseases were thought to be a result of exposure to 'miasma', or 'bad air', which was how the idea that 'fresh air' was good for you came about. And the common cold was considered to be the result of being exposed to cold air and dampness, an extension of the miasma theory.

We now know, though, that the cold is caused by a virus, which is passed from person to person by close proximity, sneezing and so on. So where did the idea that you could 'catch cold' from cold or wet weather come from? The reality seems likely to be a total inverse of the theory. In cold weather, people would come in out of the cold and huddle together near a fire or stove – sources of heat tended to be concentrated back then. And it was that clustering together in the nice warm temperatures that viruses and bacteria love that was the reason the disease was caught, rather than being out in the cold and wet.

It's true that if the body is chilled for too long it will make us more susceptible when a bug strikes, and in extreme cases can bring on exposure, but there is no truth in the idea that you can 'catch a cold' by being out in the cold and wet. It's just an old, long disproved theory that clings on in our folklore. (It doesn't help that the illness is named after the weather condition!)

LINKS:
- **Cold cures** – page 274

Chicken washing

Don't.

Many of us were brought up to wash chickens and other whole birds before cooking. Unfortunately, the splashing that accompanies the process spreads the campylobacter bacteria that are rife on poultry skins on to your skin, surfaces and even your clothes. From there they can easily be inadvertently transferred to the mouth, causing food poisoning.

The safest approach is to place the bird straight on to the cooking tray and allow the oven to deal with the bacteria. Be sure to wash your hands immediately after handling the chicken and its wrapping. In general when handling raw meat, clean any surfaces it has come into contact with using an antibacterial spray, use a separate chopping board for meat, and, as always, wash your hands after handling it.

LINKS:
- **Bacteria** – page 261

Chiropractic

**This old approach to healing through manipulation
of the spine was based on unscientific principles
and can sometimes be dangerous.**

When looking at any alternative medical treatment there are three questions to be answered. Does the explanation of the therapy make any sense? Is there good scientific evidence for it working? (Which could be the case even if the explanation doesn't make sense.) Is it a placebo, and if so, is it acceptable to use it for the placebo benefits?

In 2008, British science writer Simon Singh wrote a piece in the *Guardian* about chiropractic. As a result of his use of the term 'bogus', the British Chiropractic Association (BCA) sued Singh for libel, as they felt the term suggested chiropractors were intentionally misleading their clients, something Singh claimed he never intended to suggest, as the term was applied to the treatment, not the practitioners. The suit was withdrawn nearly two years later.

What the BCA did not dispute were the important issues covered in Singh's article. He pointed out that in the 1860s the founder of chiropractic therapy, Daniel David Palmer, developed the theory that the spine was involved in almost every illness because of the key role of the spinal cord in linking the brain and the body. Palmer wrote: '99% of all diseases are caused by displaced vertebrae.' Modern medicine makes it clear that this is simply not true – the original basis for chiropractic doesn't make sense, but this doesn't necessarily mean that the practice itself doesn't work.

Singh went on to point out that modern chiropractors, while not usually claiming as wide a range of potential cures as Palmer, have suggested that chiropractic can be used to treat a range of problems from childhood colic and feeding problems to ear infections, for which there is no medical evidence that manipulations of the spine can provide any benefit.

Even if we limit ourselves to treatment of back problems, there are issues. There is no doubt that appropriate manipulation by qualified physiotherapists can be beneficial, but Singh pointed out that a review of five studies showed that 'roughly half of all chiropractic patients experience temporary adverse effects' from pain to numbness, dizziness and headaches.

He also described how the risk from the 'hallmark technique of the chiropractor, known as high-velocity, low-amplitude thrust', though safe for most patients, can result in dislocations, fractures and damage to the vertebral arteries carrying blood to the brain, with the potential to cause strokes and death. Singh's co-author, Professor Ernst, has identified around 700 cases in the literature of serious complications from chiropractic manipulation.

Singh finished by saying: 'if spinal manipulation were a drug with such serious adverse effects and so little demonstrable benefit, then it would almost certainly have been taken off the market.'

There is no doubt that some patients swear by chiropractic, but such anecdotal evidence is not useful scientific guidance on whether or not a treatment is worth using, which is our concern in *Science for Life*. Given the evidence, it seems safest to stick to physiotherapy for back problems, and certainly not to consider chiropractic for non-back-related issues.

LINKS:
- **Osteopathy** – page 308
- **Placebo effect** – page 315

Cold cures

Getting a cold is a pain, and we'd all like to be able to make it go away. 'Cold and flu' remedies can relieve the symptoms, but beware claims of curing the common cold.

Sniffles and blocked noses, coughing and sneezing, generally feeling under the weather – it's time for the common cold to make an appearance once more. There are many products on the market that claim to relieve the symptoms, make the cold go away more quickly or even to cure it. But how effective are they?

There is no doubt that standard, off the shelf 'cold and flu' remedies can help suppress the worst of the symptoms. They typically contain paracetamol to relieve aches and pains, caffeine to help you feel more with it and phenylephrine or a similar decongestant, which may help you breathe more easily. At night-time, there are more heavy-duty mixes, like Night Nurse, formulated to help you sleep. This uses paracetamol again, but adds in dextromethorphan, a cough suppressant, and promethazine, a sedating antihistamine that helps clear the nose and makes you drowsy.

Provided they are only taken for a few days and are limited to the recommended doses, these remedies are effective and safe. But there are plenty of alternatives on sale that don't help. It's worth saying also that antibiotics have no effect at all on the cold, as it is caused by a virus, and viruses are not influenced by antibiotics.

Luckily, most cold remedies have been subject to the gold standard of medical trial reporting, the Cochrane reviews. The two most popular cold cures, Echinacea and vitamin C, are roundly wiped out. Neither has any benefits recorded for combatting the cold virus in any properly administered and controlled trial. Echinacea is a group of plants in the daisy family that has had a range of medical benefits ascribed to it, especially the ability to cure colds, while vitamin C (see main article) is, of course, an essential vitamin, but has also often been considered to have unproven benefits.

Two remedies lie in the hinterland. These are garlic and Chinese medicinal herbs. Both have anecdotal reports of success, but the trouble is that colds go away after a few days anyway, so it is impossible without a controlled trial to know whether a remedy has any effect. Add in the subjective effect that we generally feel better after taking a remedy even if it has no benefit at all – and a lot of the problem with colds is how we feel – and we can't learn anything from anecdotes. As yet, Cochrane reports that there haven't been enough randomised controlled trials to comment on these products' effectiveness as a cold cure. Take garlic by all means – it won't do any harm – but some traditional medicinal herbs can have unpleasant side effects or dangerous reactions with prescription medicine, so it is unlikely that the potential benefits of a cold cure would outweigh the risk.

One remedy with some positive results is zinc. There is some evidence that zinc supplements administered within 24 hours of onset of the symptoms can reduce the duration and severity of the cold in healthy people. However, the Cochrane report is set about with warnings that the results could have been subject to bias, and should not be taken as definitive. The other problem is that some zinc treatments have side effects of a loss of the sense of smell, and it's arguable that it would be better to suffer the cold than risk the side effect until zinc has been better researched.

Overall, then, the message for cold sufferers is to control the symptoms and get on with life. It's not thrilling to have a cold, but it's short-lived and harmless.

LINKS:

- **Catching a cold** – page 270
- **Painkillers** – page 310
- **Vitamin C** – page 114

Colonic irrigation

The use of an enema, sometimes featuring an active ingredient like coffee, to cleanse the body of toxins may be popular with celebrities, but it has no health benefits.

Enemas, which involve having a pipe placed into the rectum through which fluids are introduced and removed, do have a place in medical treatment. Colonic irrigation makes use of this approach, sometimes just using water and at other times using anything from coffee to a herbal tincture to 'remove toxins' that are said to be the cause of the various ailments being treated.

The problem here is that the body has perfectly good mechanisms for removing toxic substances. We are not generally loaded with 'toxins', and there is no such thing as a detox. There is no evidence of actual benefit from colonic irrigation, and it carries risks from potential excess removal of fluids or even in some cases a perforation of the colon. An enema is a potentially dangerous procedure that should only be handled by a professional, and there is every reason to avoid having one in these circumstances.

LINKS:

- Toxins – see **Detox** – page 34

Cough medicine

**There's a natural tendency to reach for the
cough medicine when we get a cough – but
for all their fancy formulation, there is limited
benefit from over-the-counter treatments.**

Coughs are irritating and difficult to control because they are a reflex action where the system is trying to clear either irritants (smoke, for instance) or the gunky mucus that can build up in the throat. When an irritant gets into your throat, the throat becomes inflamed, producing a dry, tickly cough, while mucus produces the chesty cough (though confusingly the mucus can be in the throat rather than in the lungs).

Commercial cough medicines tend to be either decongestants and expectorants for chesty coughs, which help break down mucus or make the mucus easier to bring up, respectively, or cough suppressants for tickly coughs, which provide a degree of soothing for the inflammation and hence reduce a tendency to cough. One of the earliest cough suppressants was heroin, though this has long been discontinued. All the evidence is that commercial cough medicines have limited advantage over a homemade honey and lemon treatment, which like them will soothe in the short term, but equally does not have long-term benefits.

There really is little better advice than to use the solution that works best for you – this is primarily about how you feel – and to make sure that you contact your doctor if the cough persists for more than three weeks.

D

• • • • • • • •

Deep Vein Thrombosis (DVT)

**When we are seated for long periods without
much movement, blood clots can form,
leading to deep vein thrombosis. Best known
as a result of long flights, this problem, which
can be fatal, can usually be prevented.**

Studies have shown that deep vein thrombosis (DVT) can hit as many as 1 per cent of the population of a developed world country per year. Although many of these occurrences clear harmlessly, some will be fatal. A study at Ashford Hospital, which serves Heathrow Airport, showed that at least one passenger a month died within minutes of arriving at the airport, and deaths throughout the year (including those whose symptoms develop later) could be as high as 2,000.

Although DVT can occur in any situation where a person is largely immobile, especially with pressure on the legs, the highest risk outside of hospitalisation (where a lot of effort is now put into avoiding DVT) is on a long-haul flight, particularly in a tightly packed economy seat. In a study of those who are susceptible to DVT and are travelling long-haul, it was found that 10 per cent developed clots, though many of these dissolved harmlessly.

Risk factors include being over 40 (the older after that age, the higher the risk), being pregnant and being on the contraceptive pill. Serious medical conditions like heart disease and cancer also increase the risk.

Planes are a particular problem in part because of the strange environment. Both the low air pressure (typically the equivalent of being around 8,000 feet up a mountain) and the very dry air, leading to dehydration, are contributory factors – and it doesn't help that on some planes the very tight seat pitch means that tall people are crammed up against the seat in front.

A simple aid is to keep hydration levels up. Don't drink alcohol on the flight, but drink plenty of water. The other simple action is to move as much as possible. Stretch and move your body parts, especially those in contact with the seat like the undersides of your legs. Get up and move around as often as is possible.

A simple way to reduce the risk is to take the same approach they do in hospitals and use support stockings. They aren't fashionable, but they work and they will be out of sight – so why not? Most chemists and travel shops now provide them.

Rather more controversial is the use of low doses of aspirin to thin the blood and reduce the risk of clotting. Half an aspirin a day for three days before the flight could have a positive effect, but don't take any medication without consulting your doctor on what is best in your particular personal circumstances.

Remember that planes aren't the only way to get DVT. It can strike as a result of long stays in bed, long car journeys or just spending all of a long day in front of a computer. In many circumstances you have more control over your surroundings than is the case on a plane. Make sure you get up and move around on a regular basis if at all possible.

LINKS:
- **Hydration** –page 63

Deodorant

There are regular scares doing the rounds on the internet that deodorants are bad for us, but there is no good evidence to support this.

Every now and then social media will be filled with re-posted messages about the dangers of deodorants and how they can cause cancer. There is no evidence to back up this claim, but the origin of the myth may be a misinterpretation of a medical instruction.

Broadly, the concerns about deodorants come in two forms. One is that by stopping us from sweating there is a build-up of toxins which can be harmful. There is no evidence of this, and sweating is not a major mechanism for the removal of toxins. The other concern is that there are substances in the deodorant that can cause harm – again the suggestion is baseless. Billions of people have used deodorants for many decades. The earliest, Mum, dates back to the 19th century, while the modern antiperspirant deodorant has been around since the 1940s. There would be better evidence by now if there were any obvious detriment.

The most likely source for these scare stories is that some deodorants contain metallic salts that can cause reflections when taking X-rays. For this reason, patients having breast X-rays are sometimes requested not to use a deodorant, and this is likely to have resulted in 2 and 2 being put together to make 5.

Drinkable sunscreen

Despite claims you may have read in national newspapers, there is no evidence that drinking a product will provide you with sun protection.

In May 2014 the British newspapers were flooded with articles about a new, miracle product. The *Daily Mail* took the lead, splashing the headline: 'World's first DRINKABLE sun cream goes on sale – and just a teaspoon will offer three hours' protection.' According to the *Mail*, the product 'works by molecules vibrating on the skin, cancelling UVA and UVA [sic] rays'. (The *Mail* clearly meant to say 'cancelling UVA and *UVB* rays'.)

This is a bizarre claim. The *Mail* tells us that the product's developer says: 'If 2mls are ingested an hour before sun exposure, the frequencies that have been imprinted on the water will vibrate on your skin in such a way as to cancel approximately 97% of the UVA and UVB rays before they even hit your skin.'

There are serious problems with this explanation. The suggestion is that somehow the frequencies 'imprinted on the liquid' can cancel out light the way noise-cancelling headphones cancel out noise. If this were possible, the military would be rushing out to buy this product for their planes as 'cancelling out light' would make them invisible. But in fact light is nothing like sound – you can't cancel it out with a vibration, even if something you drink could make your skin vibrate with a particular frequency – which it can't.

The real concern is that people will use this product and then undertake dangerous levels of sun exposure – and a particular concern is that this would seem ideal for children. There's no worse job when arriving on a beach than having to coat your children, who want to be running around, in sunscreen. Imagine how attractive the idea is of just being able to give them a drink and they are protected. But should parents do this, they will be exposing delicate skin to the sun's rays without protection, which can result in very serious outcomes.

Of course, scientists are coming up with new treatments and products all the time – but when the description of how a product works is one that bears no resemblance to known science, when the product has not been tested by any authorities for safety, and when the result of it not working could have very serious health implications, it is extremely irresponsible of newspapers to cover it in this way.

There is nothing you can drink that will protect you from the sun.

LINKS:
- **Sunburn** – page 320

E

· · · · · · · ·

Ear candles

An ear candle involves inserting an object into that delicate sensory instrument, the ear, and setting light to that object. What could possibly go wrong?

Ear candles are regularly promoted by alternative health suppliers and have been given a surprising amount of support by the media, but do they have any real benefit?

The idea is that one of these special wax candles (don't try this with the common or garden household variety) is inserted into the ear and lit. It is supposed to produce cleaning vapours that penetrate the inner ear and remove impurities and wax, drawing them out by the suction that is allegedly created by lighting the candle. Enthusiasts point out that they obviously work, as when you open up the remains of the candle it contains a waxy substance reminiscent of the earwax that we find in our ears.

Unfortunately, there are a few problems with this explanation. Firstly, that apparent waxy build-up in the heart of the candle happens whether or not it is in your ear. The substance isn't earwax, it's part of the candle. And under proper testing, there is no evidence of any suction being produced – the candles simply don't do what they are supposed to do.

What definitely has been established is that ear candles can hurt people when they get hot wax into the ear. Not only do the candles provide no benefit, there is the potential for them to do harm.

Earwax itself is a useful substance that cleans and protects the inside of the ear. It isn't something we generally want to get rid of. But in excess, it can produce earache or hearing loss. If you have earwax problems it is best to make an appointment with the nurse at your local surgery rather than attempting to remove it yourself.

Ear pops

We've all experienced that sensation when our ears need to pop – and it can become quite uncomfortable. What's the best way to deal with it?

In the old days of plane travel, the cabin crew used to come round with boiled sweets for the passengers to suck. This wasn't an early form of inflight catering, but rather an attempt to deal with an irritating problem that is most common in flight – popping ears.

The feeling of pressure behind the ears with the accompanying need for something to 'pop' is a result of changes in air pressure. If, for any reason, the air pressure around you drops, any pockets of air inside your body will expand. There is a link between your ears and your nose and mouth called the Eustachian tube, which twists and turns and is particularly good at trapping an air pocket that can cause discomfort as it expands. Surprisingly small changes in pressure can be enough to have this effect – just driving up a long steep hill in a car will usually trigger it.

The result is an uncomfortable popping of your eardrum as the air pressure equalizes. Usually it is enough to force a yawn or to swallow repeatedly, while some people can consciously exert a muscle that causes a little click in the ear and releases some pressure. But if these approaches fail, the traditional approach is the Valsalva manoeuvre. This sounds like a martial arts move, but it is simply closing your mouth, holding your nose closed with two fingers and gently blowing down your nose.

The problem can be harder to clear if you have a nasal or ear infection, or rhinitis. A decongestant medication can help in these circumstances. The experience is also usually worse for children, who have narrower Eustachian tubes, and in the extreme can lead to burst eardrums. Although modern planes tend not to have quite as large a drop in cabin pressure as the earlier ones, it is probably worth thinking twice about taking children on a flight if they are suffering from head-based infections. If in doubt, check with your doctor.

Eyesight

**Our sight is one of our most important senses, and
many of us need some correction to our eyesight,
but what do terms like 'short-sighted' really mean?**

Our eyes are our windows on the world, so it's only sensible to have some idea of the science behind their working, especially if we need a little help seeing at a distance clearly or reading small print.

Let's imagine you are looking at something specific – a word in this book. The centre part of your eye that looks black in normal light (the pupil) is a clear lens, and like all lenses, its job is to take light and focus it. In this case, the lens captures light that has reflected off the page and scattered in different directions, focusing it on a small area at the back of your eye. This back part, called the retina, has a huge array of sensors, rather like a digital camera. If the lens does its job properly, all the light coming from, say this letter A will be focused together on the same area of the retina. But if the lens in your eye isn't quite right, the image of the letter will spread out and get fuzzy.

When close-up things are fuzzy, you are described as being long-sighted. If you imagine the various rays of light from that letter A coming through the lens, instead of focusing on the back of your eye, they would have focused at some point behind the back of your eye – you would need a longer eyeball for the focusing to work. So a type of lens (a convex lens) is put in front of your eye that pulls the focusing point forward, bringing it on to the retina.

Short-sighted people have the opposite problem. When they look at something at a distance, like the letter A on an optician's chart, the image is focused at a point in the eye *before* reaching the retina, so it has got fuzzy by the time it hits the sensors. In this case, another type of lens, a concave lens, is used to push the focusing point backwards. Because the two problems are the result of trying to look at things at very different distances it is possible, especially with older eyes, to suffer from both conditions, requiring bifocals.

The third major type of correction you might need is for astigmatism. Here the focus varies depending on the direction the light is coming from – the lens in the eye isn't totally symmetrical and requires a lens shaped differently left-to-right from top-to-bottom (called a 'toric' lens because it is like a slice off the edge of a torus, the mathematical name for the shape of a ring doughnut).

The numbers on your optician's prescription will refer to the strength of the lens in diopters, with a negative number indicating a correction for short sightedness and a positive number for long sightedness. There may also be a second strength for the variation required for astigmatism and an angle that shows the 'tilt' of the correction to deal with the astigmatism.

F

• • • • • • • •

Five-second rule

**Everyone has probably been told at some point
when food is dropped on the floor that it's okay to
eat if you pick it up in under five seconds. But is it?**

It's rare for food to be dropped on the ground without someone
invoking the five-second rule – the idea that if you drop a piece
of food, you should be okay as long as you pick it up within five
seconds. But is this true or an old wives' tale?

Apparently, this approach dates all the way back to the time of
Genghis Khan, though back then, when people were less fussy, it
was the twelve hour rule. A US high school student, on a summer
course at a local university, took a more modern scientific approach
to the rule and made some interesting discoveries.

When Jillian Clarke took swabs from floors at the university,
including areas with a high footfall, she was surprised to discover
that the floors were remarkably clear of bacteria. The PhD students
helping her couldn't even find countable numbers. They did discover, though, perhaps not surprisingly, that people are less likely
to pick up and eat broccoli or cauliflower from the floor than sweets
or biscuits. (In fact, this surely applies whether or not the foodstuff
is on the floor.)

Perhaps the most important finding though was that when a
surface was inoculated with E. coli bacteria, foodstuffs did pick up
the bacteria in under five seconds – so in that sense the rule fails.
It's really just common sense that the five-second rule is rubbish.
Bacteria don't walk to the food and jump aboard, so taking some
time to achieve it. As the food hits a surface, it rubs some bacteria
off. There is no time short enough to avoid it.

Of course, not all bacteria are bad for us. In fact, we get a lot of benefit from the bacteria in our gut. But any surface could be contaminated with the infectious kind of bug, plus viruses, fungi, faecal matter and a whole range of toxins that no one really wants to swallow.

What doesn't seem to have been checked was whether there was a difference between a dropped object with a hard shiny surface and a sticky or wet object. Without suitable research to back it up, it seems likely that sticky and wet objects are more likely to pick up contamination and we are still likely to give a resistive dry object, like a tablet or an M&M, a quick wipe and swallow it. But what remains certain is that the five-second rule isn't serious guidance.

LINKS:
• Good bacteria – see **Probiotics** – page 80

H

. . . .

Hand washing

We all know it's important to wash our hands before touching food, but a good hand wash is more about technique than using a particular product.

Choosing a way of getting clean hands can seem a major purchasing decision these days, with options ranging from antibacterial alcohol gels to good old-fashioned soap and water.

There is no doubt that there are circumstances where an antibacterial product, like those used in dispensers in hospitals, is a good thing. There aren't always the facilities to wash your hands properly, and in such circumstances an alcohol-based gel is worth using. But should we be using them all the time, and if we do wash, should we use ordinary soap or go for an antibacterial handwash?

All the evidence is that alcohol gels are second best – worth using where there is no other choice, but not as good as traditional hand washing. Used properly, traditional soap and water can remove a good 99 per cent of bacterial and other unwanted matter from the skin, where a gel is only likely to come in at around 90 per cent.

When it comes to different types of handwash, there is some evidence that it is best to avoid antibacterial products. This is partly because studies have shown that some antibacterial agents can prevent thyroid hormone from functioning normally in animals (though results have not been established in humans). But mostly it is because it is strongly suggested that heavy use of antibacterial washes can increase the resistance of bacteria to antibiotics, which is already a major problem for health and will increasingly be so. (Antibacterial washes are still important in healthcare settings, but not in the home.)

The good news is that perfectly ordinary soap or liquid deter-gent is excellent at removing unwanted material, but most of us don't wash our hands effectively. The four essentials are:

- Use hot water if at all possible.
- Work the soap and water mix around all of your hands from palms to fingertips and between the fingers. Don't just rub around the centre section.
- Wash for long enough – most of us are far too quick.
- Don't put bacteria back on with a dirty towel.

Almost all of us fail on not washing for long enough. The recom-mendation is that you continue washing for the time it takes you to sing through 'Happy Birthday to You' twice (in your head – no need to do it aloud). This feels a painfully long time, but is what you need to be truly effective. If you can't bring yourself to do this every time, at least attempt it before handling food.

The duration of the wash is worth far more than any fancy hand-wash product and is the single thing you can change that is most likely to increase the quality of your hand washing.

LINKS:
- **Antibiotics** – page 258
- **Bacteria** – page 261

Hangovers

**There is no doubt that hangovers are real.
But can they be avoided or cured? And is a
chemical known as acetaldehyde the culprit?**

The hangover is an unpleasant, self-imposed malady. But can it be avoided or cured? It would help if we knew the cause, and for a number of years a prime candidate has been a substance called acetaldehyde. The name is derived from 'acetic alcohol dehydrogenated' because it's a substance that only differs from ethanol, the compound that makes drinks alcoholic, by having one less hydrogen atom.

Ethanol is a naturally occurring poison that the body has inbuilt mechanisms to counter. In the liver, an enzyme turns alcohol into acetaldehyde, which is even more toxic than alcohol, and a second then makes the acetaldehyde into acetic acid. If this process doesn't work quickly enough, the system allows acetaldehyde to build up, something that can happen if we consume alcohol too quickly.

The result is feeling extremely nauseous, having powerful headaches and vomiting. This is a process that is triggered intentionally by the anti-drinking treatment Antabuse, which blocks the enzyme that breaks down acetaldehyde, so a patient on the treatment suffers the horrible effects of acetaldehyde poisoning if they take alcohol.

While the symptoms of acetaldehyde poisoning are similar to those of a hangover, they are more acute and happen earlier in the process – so while it is likely that acetaldehyde is responsible for much of the suffering of binge drinkers at the end of their night out, it has a smaller contribution to a traditional morning-after hangover. This is just as well, as the best way to prevent acetaldehyde poisoning is to drink less alcohol, and spread your drinking over a longer period of time, and it's usually too late by the time you get a hangover to do anything about this.

One study does suggest that there are ways to help the enzyme that renders acetaldehyde harmless. Media coverage said that the study showed the fizzy drink Sprite made a good hangover cure, as it speeded up the action of the enzyme. However, this Chinese study didn't explicitly name the product (though the '*xue bi*' in the study may be Sprite), and it didn't test the impact on hangovers, just on the enzyme in a lab. As we see time and again, just because a substance interacts with another in a test tube does not mean that consuming that substance will have the same effect in the body. The same study also found that some substances, notably green tea, had the negative effect of slowing down the enzyme (though again this may well not happen in the body).

One major contributory factor to a hangover is simply not getting enough sleep. Another problem is dehydration. Alcohol is a diuretic, increasing urine production and leaving the drinker dry-mouthed, nauseated and with a thumping head. The dehydration aspect means the now popular idea of drinking plenty of water at the end of the night and in the morning is a good move and will help reduce the impact of an evening's drinking. But sports drinks and other 'hydration' products are no better than water, and coffee, itself a mild diuretic, is not the ideal solution.

What's more, the enzymes used to turn alcohol into acetaldehyde and then to acetic acid have other important roles in the metabolism. While they are busy rescuing your body from an alcohol attack, they are not contributing enough to their other roles, often resulting in a drop in glucose levels in the brain, producing tiredness, mood swings and lack of concentration. A good carbohydrate load might help a little, and though the traditional fry-up is probably not a great idea, managing to eat some toast and fruit juice should help those blood sugar levels.

Finally, most alcoholic drinks contain a whole bundle of other natural chemicals, like aldehydes and esters, which themselves can have deleterious effects on the body. So avoiding drinks with a higher load of the secondary chemicals – the worst culprits are whisky, brandy and fortified drinks – can also reduce the impact.

There is no miracle cure for a hangover, other than not drinking as much alcohol. As a *British Medical Journal* survey of randomised trials concluded: 'No compelling evidence exists to suggest that any conventional or complementary intervention is effective for preventing or treating alcohol hangover. The most effective way to avoid the symptoms of alcohol induced hangover is to practice abstinence or moderation.'

LINKS:
- **Alcohol** – page 5
- **Hydration** – page 63

Herbal medicine

Herbs have been used in all cultures as medicines and some have real medical benefits. But care has to be taken with herbal cures as they are poorly regulated.

In historical times, herbs played a major part in our medical treatment, though as early as the 17th century, when herbalist Nicholas Culpeper published his *Complete Herbal*, there was a debate between the proponents of herbal and chemical medicine. With a modern understanding of the active ingredients in herbs we can see that the distinction is artificial. Herbs are an effective source of chemicals that have an active effect on the human body, though often this effect is improved if the chemical agent is refined and modified.

So, for instance, willow bark, containing a chemical called

salicylic acid, has been used to help reduce pains and fevers for at least 4,000 years. Unfortunately salicylic acid causes sharp stomach pains and can cause internal bleeding – so that relief came with a price. At the end of the 19th century, the German pharmaceutical company Bayer produced a painkiller called aspirin, which modified the active ingredient to the much less aggressive acetylsalicylic acid. The active ingredient of medical herbs has been repeatedly isolated and improved. Quinine for malaria, atropine and digitalis for the heart, for instance, all originated with herbal cures, but are much better and safer in their pure chemical form.

Herbal medicine has a strong presence in China, and there are herbal medicine stores on many of our high streets. Unfortunately, current herbal medicine still uses theories with no basis (such as employing a herb with the same shape as an organ of the body to treat that organ) and has rarely advanced. There are also well-documented dangers in taking herbal medicines at the same time as other medications. It is always worth checking with your doctor first.

The popular herbal treatment St John's Wort, for instance, which has proved effective in helping with mild depression, has as many possible side effects as any prescription drug, from gastrointestinal pain to dizziness. It can cause serious problems with anti-HIV, anti-cancer drugs and contraceptive pills; plus it generally reduces the effectiveness of the transport mechanism carrying drugs into the bloodstream. Because of this it has been banned in France. You can't assume that because a treatment is herbal it is safe in all circumstances.

Another significant problem with herbal treatments is that a herb rarely contains only one active ingredient – and it is almost impossible to give a well-controlled dose, as the amount of the chemical in a herb will vary from plant to plant. Where a prescription medicine will have isolated the useful chemical, in a herb you could be exposing yourself to a whole range of potentially unpleasant substances to get the beneficial one. But more worrying still is the reality that there is very little regulation of the quality of herbal medicines.

A major US study in 2013 analysed 44 processed herbal products and 50 medicinal herbs from a range of suppliers. The results were shocking. Less than half of the products actually contained the labelled main ingredient. Nearly 60 per cent contained herbal species that weren't listed on the ingredients. A third also contained fillers and contaminants that weren't on the label. In three of the companies tested, *none* of the products contained what they said they did. In other studies, herbal medicines have been found to be contaminated with dangerous heavy metals and other extremely poisonous substances.

So, while there are certainly benefits to some herbal remedies, the lack of a proper regulation and of a testing regime paralleling that used on conventional medication means that there will always be significant risks attached.

Homeopathy

Homeopathy is one of the most popular alternative medical approaches, favoured by Prince Charles and millions of others. But scientifically it is highly dubious.

Until recently, homeopathy had support from parts of the NHS in the UK, but it is being side-lined. However, it remains very popular for self-medication. It is particularly widespread in France and Belgium, where around a third and a half of the population respectively make use of it.

When looking at any alternative medical treatment there are three questions to be answered. Does the explanation of the therapy make any sense? Is there good scientific evidence for it working? (Which could be the case even if the explanation doesn't make sense.) And is it a placebo, and if so, is it acceptable to use it for the placebo benefits?

The reasoning behind homeopathy is that a small amount of a (usually poisonous) substance that causes particular symptoms when consumed should be used to treat an illness that causes the same symptoms. In practice, the 'active' substance goes through so many dilutions that it is incredibly unlikely that there will be a single molecule of the substance left in the final solution, which is usually dripped on to a sugar pill. A typical dilution found on the shelves is '30C'. This means the original substance has been diluted by a factor of 100, 30 times in row. (This makes the chance of a single molecule of the active ingredient being left in the medicine a million trillion trillion times less likely than you winning the UK lottery jackpot with a single ticket.) During the dilution, the solution is shaken vigorously, a process known as 'succussion', which is said to improve the effectiveness of the treatment.

Supporters of homeopathy are aware of the dilution problem, but some believe that water can have a memory that enables it to act as if the substance were still present – this does seem to be clutching at straws. To increase the lack of believability, homeopathy insists that the more dilute the substance, the more powerful it is. It may be intentional that homeopathic remedies do nothing. The treatment was developed at the end of the 18th century by German doctor Samuel Hahnemann. He was aware that many of the treatments of the time did more harm than good, and it is arguable that doing nothing but making the patient feel better was the best option available as many others (bleeding, for instance) made the patient weaker and less able to recover.

The whole concept that the water could have memory is difficult to reconcile with any known science or even logic. As the Australian Council Against Health Fraud has pointed out, it is strange that the water's memory is so selective. How does it know to remember the homeopathic cure, but not the various bladders it has passed through in the past and all the other chemicals that have been in it and then diluted out of it on its way to the final place of use?

Although homeopathy stretches credulity to its limits, in the 1980s, the prestigious journal *Nature* published research by French scientist Jacques Benveniste that suggested there was an effect from an ultra-diluted solution, where all the original substance was removed. However, there was a problem with the original work, and it echoes one of the common pieces of anecdotal evidence supporting homeopathy.

Lots of people say that they have had benefits from homeopathy, but the anecdotal success could be down to the placebo effect. If we think something is helping us, we usually will feel better, whether or not there is any actual effect from the treatment itself. However, when I first talked to a homeopathic enthusiast, she was convinced I had it wrong. 'How can that be,' she said, 'because homeopathic treatment also worked on my horse. It doesn't know it is being treated, so it can't be a placebo.'

The problem here is that the person giving the treatment *does* know about it. The animal (or person) being treated doesn't need to be aware of the 'medicine' to pick up the positive feelings involved. What's more, when such treatment has been monitored, it is often the case that the person administering the treatment will give the animal more attention than usual, which will inevitably improve the animal's sense of wellbeing.

What the Benveniste trials reported in *Nature* – and anecdotes of homeopathic treatment – fail on, is double blind testing. The *Nature* trial was blind – the patients didn't know whether they received a homeopathic pill or a placebo, but it was not double blind – the people running the trial did know which was which. When someone says 'a homeopathic treatment made me feel better,' the anecdote is not even about a blind trial. Animal treatment, just like the *Nature* trial, is blind but not double blind. The person administering the treatment knows.

When the *Nature* trial was repeated (by the same researcher, in identical conditions) with the researcher unaware which was the homeopathic treatment and which the placebo, there was no effect. The apparent ability of the homeopathic treatment to work had disappeared. As is often the case with alternative medicines, there have been a fair number of trials that have had positive results, which sounds encouraging. But all serious analyses taking the results of many trials and combining them, have come to the conclusion that homeopathy is nothing more than the placebo effect. This has proved the case both with animal trials and humans.

The good news is that, unlike some alternative medicines, homeopathic treatments are truly harmless. (This is sometimes demonstrated by people taking massive overdoses of homeopathic pills.) They can make those who believe them feel better in themselves, so may be a good remedy for minor aches and pains, where the alternative is to do nothing – although there is always the concern about the ethics of treating someone with a lie, and the cost of homeopathic remedies far exceeds that of the sugar pills that are received, suggesting that anyone wanting to take this approach would be better off providing their own sugar pills. The source of the homeopathic flu treatment Oscillococcium for a whole year is based on extracts taken from a single duck. In final form these sell for around £12 million worldwide. Quite a mark-up.

However, there is a dark side. Some homeopaths claim they can treat serious medical conditions like malaria or cancer, or offer to travel to a disaster zone to help those suffering from dysentery, cholera, etc. Others have dissuaded patients from using conventional medicine – in a study during the MMR fiasco, around half of homeopaths examined attempted to persuade parents to avoid the MMR vaccine, and in another study most homeopaths advised using dangerously ineffective homeopathic antimalarial treatments rather than essential drugs. Any case of a serious illness where someone takes a homeopathic remedy, and because of this fails to take functional medication, means that the homeopathic treatment puts the patient at risk – and in those circumstances it should be avoided at all cost.

LINKS:
- Double blind testing – see **Blind trials** – page 262
- **MMR** – page 305
- **Placebo effect** – page 315

Hyperactivity and sugar

**Everyone knows that sugar makes
children hyperactive. But is there
any science behind this idea?**

There's an episode in *The Simpsons* where Lisa and Bart eat European chocolate and start literally bouncing off the walls because it has a 'higher sugar content than they're used to'. It's accepted pretty much as fact that giving children sugary sweets and drinks makes them hyperactive – and yet there is no basis for believing this.

At least twelve double blind studies have been undertaken where children were given sugar or a placebo (for once, not a sugar pill). In none of them was there any evidence of change of behaviour as a result of the sugar, not even in children diagnosed with ADHD or considered to be highly sensitive to sugar by their parents.

There was some evidence that much of the effect was down to expectation. If parents were told their children had just been given a good stiff dose of sugar, they then believed that their kids were behaving more rowdily than normal.

It is possible that this myth dates back to a time when sugary treats were less common than they are today. This meant that just being given something sweet was a cause of excitement – and one that was often associated with the extra stimulation of parties. The children's natural exuberance at these events was interpreted as being caused by the sugar. It's something we often see– it's assumed that when two events happen at the same time (the children have sugar and they get excited) then one causes the other.

To make matters worse, the energy boost from the sugar might help the children keep partying longer. But it doesn't mean that it caused the uncontrolled excitement in the first place.

Of course this doesn't mean that high-sugar diets are a good thing. Most children consume far too much sugar, damaging teeth and seriously increasing the risk of obesity, heart disease and diabetes. It's important to keep an eye on your kids' sugar intake. But next time a doting grandparent gives them a bag of sweets, don't panic. The children won't start bouncing off the walls.

LINKS:
- **Sugar** – page 102

I

• • • • • • • •

Immune system

**Our immune systems provide vital protection
against attack, but be wary of claims that a
particular substance can 'boost' your immune
system – leave that to vaccinations.**

You don't have to look far through a health supplies store – or
through the pages of many magazines – to find a food or supplement that it is claimed will help boost your immune system. There
is a way to do this, but it is a very specific approach, as we will
discover.

Your immune system is not a single part of your body but rather
a vast network comprising physical barriers like your skin, white
blood cells, various different organs and a whole range of complex
chemicals with literally thousands of different roles. 'Boosting' it
by simply eating something is a bit like hoping to redecorate your
house by throwing a capsule of paint at the wall.

It might seem that all the claims and counterclaims made for
immune system boosters are very scientific, and they will often state
that a particular product is *scientifically proven* to boost the immune
system – but when it comes down to detail, there is never good
scientific data to back this up. It's one thing to make the claim. I
could say that eating oysters will double your strength. I could say
it's scientifically proven, and I know several people for whom it has
worked. But in reality it's not true, and I have nothing to back up
the claim other than a few stories that I've invented.

What often happens is that a legitimate study is used to draw
inappropriate conclusions about the importance of a particular
food to the immune system. A while ago, a report was published

showing that when banana was introduced to the abdominal cavity of rodents it resulted in the body producing a substance called TNF. This is the factual part. But this was interpreted as meaning that bananas contain TNF (they don't), so eating bananas will introduce this immune system booster, which could cure cancer. But even if bananas did have TNF (they don't), you wouldn't absorb it by eating them. And TNF doesn't cure cancer.

In reality, your immune system is very robust. It isn't weakened by having a disease (except special conditions like HIV/AIDS), though it can be suppressed deliberately and has to be in some circumstances, for instance by treatment to avoid rejecting transplants. Most of us don't want a boosted immune system, the immediate outcome of an overactive immune system being allergies and conditions like eczema and asthma. But this isn't a problem because there is nothing that has been shown to boost the immune system in the way that the term is used by vendors of supplements and special foods.

However, the immune system has a limited repertoire. It can only handle attackers that it knows about or that fit a recognised pattern. When something new comes along, the immune system can be helped to respond more quickly and effectively. What can be done is to introduce a harmless substance that triggers the immune system to protect the body against a real threat. Once this has been done, the body can fight off a new attacker, so the immune system has been extended. This process is known as vaccination.

LINKS:

- **Allergies** – page 254
- **Vaccination** – page 323

J

. . . .

Jet lag

**There is no doubt that jet lag exists.
However, it is not a medical condition,
but rather the symptoms of fatigue
and other contributory factors and
needs to be treated accordingly.**

It's a good idea to be suspicious of miracle cures for jet lag, as it is not a medical condition – you can no more 'cure' jet lag than you can 'cure' being tired. There are ways to alleviate the symptoms or prevent it happening as badly, but you can't cure it.

The main problem is fatigue. A long-haul flyer suffers from a combination of exposure to low pressure air, dehydration, stress and the kind of disruption of sleeping patterns familiar to shift workers.

You can overcome jet lag by simply waiting, but it takes about a day per hour of flight to fully recover, so it is sensible to minimise the impact. Start before the flight by getting a few good nights' sleep. In the 24 hours before flying, don't eat or drink heavily and stick to bland food.

As soon as you get on the plane, set your watch to the destination time and operate to that time, both in terms of sleep pattern and meals. Unfortunately, airlines don't get this and will often stick to the departure time. As much as you can, ignore them – a sleep mask is useful, and if you can't resist any meals being served, at least try to fit them as best you can to your destination times. In practice, it is best to keep food consumption down to a minimum on the flight.

As for drinking, avoid alcohol (even if it is free), but drink plenty of water. It is best not to overdo carbonated drinks (including sparkling water) as the low cabin pressure can exacerbate gas build-up in the gut.

There have been studies into the benefit of using the drug melatonin to reduce jet lag. This hormone is supposed to fool the brain into ignoring the impact of changing time zones. Research has shown that the body's levels of melatonin vary after a long flight, and there is a strong link between melatonin levels and the ability to sleep. But there are real questions over the prescribing of melatonin as a 'cure' for jet lag. According to the UK medical journal, *The Lancet*:

> [Melatonin's] apparent usefulness in alleviating the effects of jet lag may be related to some ill-defined psychotropic activity, but such effects may be undesirable, particularly if they modify daytime function [...] Melatonin is a possible inhibitor of sexual development in rats and may initiate gonadal regression in voles. It also has endocrine effects on man. The effects cannot be dismissed lightly...

If the drug works at all, it is necessary to undertake a rigid timing programme of doses around your flight time. Get the schedule wrong and the drug will make jet lag worse. There is some evidence that melatonin has an effect, even if the possible side effects are worse than the original problem. By comparison, some suggested jet lag cures are based entirely on fooling ourselves. There is a whole range of alternative therapy and exotic treatment available, but there is no good evidence that any of these approaches have benefit above the placebo effect.

LINKS:
• **Placebo effect** – page 315

M

• • • • • • • •

MMR

The MMR scare caused serious health problems for children with dangerous outbreaks of measles. The science is clear. There is no link between MMR and autism.

The MMR (combined measles, mumps and rubella) vaccination scare is probably the worst example ever of the media making a terrible mistake in its reporting of science, then failing to admit just how wrong it got things.

In 1998, one individual with no appropriate scientific qualifications, Dr Andrew Wakefield, made a claim that there was a link between autism and the MMR vaccination. Wakefield's 'study' was based on only twelve children and made no attempt at the kind of controls essential for a proper scientific investigation. Most of the children were selected *because* their parents believed that they had begun to suffer from autism as a result of having the MMR vaccination. There was no evidence whatsoever for a link provided by the 'study'. Similarly, there was nothing in the study that suggested individual vaccinations were safer than MMR – but this was what Wakefield called for as a result.

Since then, the *British Medical Journal* has come up with the conclusion that: 'Wakefield's article linking MMR vaccine and autism was fraudulent.' Investigative journalist Brian Deer showed how, before undertaking the 'study', Wakefield had been hired by a lawyer to attack MMR in a speculative lawsuit, and had been paid to try to discover evidence to support the legal case. Wakefield was also discovered to have filed a patent on a 'safer' single vaccine, which would be valuable if MMR was discredited. In 2010, Wakefield was found guilty of charges including dishonesty by the General Medical Council and was struck off.

Even today, searching for information on Wakefield online you will find plenty of material in his defence – yet this is a totally discredited man with a totally discredited theory. Unfortunately, as a result of the initial media coverage, which has never been apologised for, UK vaccinations slumped and are only now returning to safe levels. This means that there have been repeated outbreaks of measles, which is a very serious childhood disease that can and has caused extreme illness, disability and even death.

There is a very clear message. The MMR vaccine is safe. As with all vaccines, there are small risks of side effects, but their impact is tiny compared with the devastation caused by measles outbreaks. There is no reason to use individual vaccines and no evidence of any benefit from doing so. It is important to have children vaccinated, and there is no evidence whatsoever of a link to autism. This was an unfortunate medical error, possibly perpetrated for financial gain despite the harm it did to children, and magnified by the disastrous media response.

LINKS:

- **Vaccination** – page 323

O

. . . .

Obesity

**There is no doubt that being seriously overweight
presents health problems, and obesity is
an issue for all developed countries.**

There are far more people who are overweight or obese than there
once were, and this is a serious problem for any nation's health. In
March 2014, the UK's chief medical officer suggested that obesity
was now 'seen as normal in society.' But what does being 'obese'
really mean, and what can you do about it?

As is discussed in the **Body Mass Index** section, it is possible to
be labelled obese based on BMI even if you are very fit and healthy,
as muscle is more dense than fat, so the definition of obesity coming from having a BMI of 30 or greater has to be taken with a pinch
of salt. However, if you have a high BMI you will probably know
perfectly well if that can be put down to well-honed muscle, or if it
is a subject of concern.

You only have to look at people on a typical high street to be
aware that obesity is on the rise. At the time of writing, the number of adults in the UK classified as obese is rising from around a
quarter to around a third. Around 20 per cent of children in the
UK are classified as obese. More scarily still, it has been suggested
that current predictions – that half the UK population will be obese
by 2050 – significantly underestimate the reality. (The one slight
encouragement is that statistics for overweight and obese flattened
out in 2012/13, but it is too soon to say if this is a trend or a one-off
event.)

Research has made it clear that obesity causes three key problems. It increases the risk of heart disease, stroke and type 2 diabetes.

It makes everyday life harder – and more expensive for the taxpayer, as the health problems caused have a significant impact on National Health Service budgets. And it impacts our self-esteem. Although there are people who publicly claim that they are entirely happy with being 'big', no one sets out to be obese on purpose.

Governments run all kinds of campaigns to encourage us to eat more healthily, to eat less and to exercise more – and it's easy to blame fast food vendors, sweet manufacturers and bigger portion sizes. The UK chief medical officer is suggesting a 'sugar tax' to reduce sugar consumption. And all this can help. But in the end, it is down to us as individuals and families. Obesity can be tackled, but only if we have the will to personally do something about it.

LINKS:
- **BMI** – page 264

Osteopathy

**A popular alternative to chiropractic, osteopathy
also involves a degree of manipulation, but is
usually gentler and more massage-like in style.**

Osteopathy dates back to a similar point in time to chiropractic, and also involves manual handling of the back. It is frequently recommended for back treatment, but the manipulation is less extreme and has a lower risk of damage than chiropractic.

All the evidence is that osteopathy can be an effective treatment for back pain, though there is no evidence that it is any better than conventional physiotherapy, and it is recommended that any osteopathic treatment is undertaken by a qualified physiotherapist, ensuring that the therapist has the best working knowledge of the underlying structures and potential harm that can be caused by manipulation.

As is the case with chiropractors, some osteopaths claim to be able to treat a wider range of ills from colic to ear infections. Not surprisingly, there is no evidence that this is an effective treatment in any of these cases.

LINKS:
- **Chiropractic** – page 272

Painkillers

**By far the most common type of
self-medication, painkillers are generally
regarded as harmless and useful. But is
this true – and which works best?**

Until the 1970s the most common household painkiller was aspirin. This was the first large-scale, over-the-counter painkilling drug, based on a herbal treatment using the bark of the willow tree and extracts of the plant spiraea, or meadowsweet, that went back thousands of years. These folk treatments reduced pain and fever, but could cause sharp stomach pains or internal bleeding. Aspirin uses a variant of the active chemical to reduce the impact on the stomach, but still can cause problems, which is why it is rarely used now as a painkiller.

Today, the most common painkiller is paracetamol (called acetaminophen in the US). As long as you stick to the recommended dose it is safe, effective and quick acting, and we happily use it on our children in the form of liquid painkillers like Calpol, without the concerns of stomach damage that are known to accompany aspirin. Of course, paracetamol has its dark side – if the recommended dose is exceeded it is easy to overdose, which can result in liver failure and death. But taken in a controlled way, it combines safety with reliability.

There has been some indication in recent studies that long-term usage of paracetamol can also cause stomach issues, but this should not have an impact on typical use, as it only applies to those taking the drug day in, day out. For some individuals, paracetamol also has little more pain relieving impact than a placebo, but these are relatively few, and for most of us it remains a good, occasional-use painkiller.

There are two more recent over-the-counter painkillers, ibuprofen and diclofenac. Ibuprofen is particularly effective with muscle pain, but does aggravate the stomach in some users, an effect that can be increased if drinking alcohol. If necessary, ibuprofen can be taken at the same time as paracetamol to provide a more concentrated painkilling effect. Diclofenac, like ibuprofen, is a 'nonsteroidal anti-inflammatory drug' and is even more targeted on muscle pain and inflammation. It carries similar concerns to ibuprofen over stomach conditions and should be avoided in the third trimester of pregnancy.

A final painkiller worth mentioning is codeine. This is often sold in dual tablets with paracetamol, giving extra painkilling properties. Codeine is the most powerful over-the-counter painkiller as it is an opiate, related to morphine, and should be used sparingly as it is potentially addictive. Never use codeine for more than a day or two without medical advice.

No painkillers should be taken for more than a few days without consulting your doctor – all painkillers carry some risks when used long-term, and there may be an underlying condition that needs to be examined.

Panic attacks

A panic attack can be extremely scary, with symptoms that can seem like a heart attack, yet they are usually easily controlled.

Many of us suffer from a series of symptoms that can seem like the onset of a heart attack. The sufferer has breathlessness, often accompanied by numbness and tingling in the limbs. There is a tightness in the chest, clammy skin and it is often accompanied by palpitations and dizziness. It is terrifying. Yet these are the symptoms of a panic attack, a stress-related body reaction that is self-reinforcing, as the unpleasantness of the symptoms causes more stress and builds the effect.

The symptoms are caused by hyperventilation – effectively over-breathing – which can be triggered physiologically by lower than usual levels of carbon dioxide. We don't need carbon dioxide to breathe – in fact too high a level is life-threatening – but the body uses carbon dioxide as a monitor of its state of balance, and in a panic attack is misinterpreting the fall in level. The result is highly unpleasant for the sufferer, and the symptoms are sufficiently worrying that they can result in mistaken calls for medical assistance.

A plane journey makes for the ideal conditions for a panic attack, as it combines a powerful stressor in the experience of flight with the lowered carbon dioxide levels that result from the relatively low cabin pressure.

If suffering a panic attack, you should try not to move more than you have to until the symptoms have receded. If you are driving, pull over. The traditional way to counter a panic attack, apart from the relief of realising that this is a harmless reaction, not a major life-threatening condition, is to push the carbon dioxide levels back up. This is done by putting a paper bag over the mouth and nose, so that the sufferer re-breathes exhaled air. As the air we breathe out has higher levels of carbon dioxide, this tends to help the body re-establish its balance, and the symptoms soon fade. If no paper bag is available, a hand cupped over the mouth and nose would usually provide similar relief.

Of late, this technique has fallen out of favour and instead panic attack sufferers are encouraged to change their breathing pattern, consciously taking steady, slow breaths to reduce hyperventilation.

It is also recommended that they try to focus on something external that is simple and non-threatening. As much as possible the mental focus should not be on the attack itself.

If you have several panic attacks it is important to get medical advice.

Paracetamol and childhood asthma

Paracetamol, the painkiller most often given to children, has been linked in some studies to childhood asthma, but there is as yet no evidence that paracetamol causes the asthma.

Paracetamol (acetaminophen in America) is a widely used painkiller. Like all such drugs, it should be used only when necessary, and as is made clear in the **Painkillers** section, paracetamol overdoses are extremely dangerous. But taken appropriately, paracetamol has fewer side effects than most other painkillers and is taken millions of times a day without incident. Anyone with children is likely to have come across low dose paracetamol in liquid form in popular childhood painkillers. However, there have been significant concerns linking paracetamol to childhood asthma. To see why these serious scientific studies are easy to misunderstand, it's important to understand the difference between correlation and causality.

Correlation means that two things are linked in some way. When one goes up, for instance, so might the other. But correlation does *not* mean that one thing causes the other. Causality is a stronger linkage, where one thing does actually cause the other. The birth rate in the UK for several years after the Second World War was correlated to the import of bananas. In a year where more bananas were imported, more children were born. Fewer bananas and there were also fewer children. But no one would deduce that the bananas caused the pregnancies. It is more likely that a third factor, such as disposable income, influenced both. You could imagine a cause the other way round – if the pregnancy rate caused the change in banana imports – as it's just possible that pregnant women eat more bananas, though it is unlikely.

Bearing in mind that just because A and B rise and fall together doesn't mean that A causes B, several studies have shown that children who were given liquid paracetamol at least once a month had a significant increase in reporting asthma symptoms. However, none of the studies provided evidence that the paracetamol *caused* the asthma symptoms – only that the symptoms were more common in those who took paracetamol regularly.

What we need to bear in mind is that if there is a causal link, it is more likely to be the other way round. It seems reasonable, indeed, that people with children who get asthma symptoms may be more likely to treat them with a painkiller.

At the moment, on the balance of evidence, the benefits that millions of children and their parents get from careful use of paracetamol outweighs a small possibility of a causal link. Eventually more certain evidence, with the ability to identify causes, will be gathered – but this takes a lot more time and complexity of studies than simply identifying a correlation, as has been done so far. Based on the data currently available, the medical regulatory authorities agree that paracetamol should continue to be given to children.

LINKS:
- **Painkillers** – page 310

Placebo effect

The placebo effect is a positive medical result, for example the reduction of pain, caused by a patient's own body in the response to a belief that a treatment is doing some good.

British doctor John Haygarth discovered the placebo effect in 1799. Haygarth was suspicious of quack medical devices known as 'tractors', metal rods that were supposed to have the powers to 'extract' pain from the body. Patients swore by their effectiveness, but Haygarth was doubtful. The tractors were supposed to work because they were made of a special, expensive, secret alloy. Haygarth had fake tractors made up of ordinary metal and arranged for a comparison of fake and real tractors, without the patients knowing which was used. As he suspected, patients using both 'real' and fake tractors claimed benefits.

The placebo effect was discovered to work, for instance, with sugar pills containing no active ingredient. If patients thought they were getting a powerful medicine, some felt better. This effect didn't work for everyone, and didn't work for all conditions. It seemed to be particularly effective for some kinds of pain relief. It was important that the patients did not know if they had the real or fake treatment, as knowing the treatment was fake meant that they were less likely to feel a positive effect. (Equally, it is possible for people who think they are taking tablets with a known side effect to feel that side effect even if they are taking a sugar pill. This is called the *nocebo* effect.) Trials comparing a drug with a placebo are known as 'blind' trials and are essential for medical testing.

Research showed that the placebo effect was not always in the patient's imagination, but that the patient's body could release natural painkillers in response to the belief that the fake medication would have effect, and these natural painkillers genuinely reduced the impact of pain.

This leads to a moral dilemma. Should a doctor lie to a patient, and allow them to feel better while having a non-existent treatment, or tell the truth, in which case the placebo is less likely to be effective (though strangely it does still work for some)? Is it acceptable to lie to help a patient to feel better?

What is clear is that it is one thing to use placebos medically, and it is another to sell a fake medical 'cure' that is a placebo, particularly if that 'cure' is extremely expensive, when the actual treatment given has no or very little value.

LINKS:

- **Blind trials** – page 262
- **Painkillers** – page 310

R

· · · · · · · ·

Reflexology

**This method of diagnosing and treating
medical problems by massaging the
feet has no medical benefits, but foot
massage provides valuable relaxation.**

The idea of reflexology, like many alternative medical treatments, came from a single individual who had an idea with no good reasoning to support it. He believed that the different parts of the body are mapped on to the soles of the feet, and that by feeling for resistance in the relevant areas, a trained practitioner could diagnose internal illnesses and then treat them by applying pressure to the point, which somehow was supposed to help restore the malfunctioning organ.

There is no medical basis for this theory, and when undergoing controlled trials, reflexologists have shown no ability to diagnose actual conditions or to treat them.

The only apparent benefits of reflexology are those of any foot massage – and as reflexology usually costs considerably more than a simple massage, if you enjoy having your feet worked on, you would be better sticking to an ordinary masseur.

LINKS:
·   **Breathing and relaxation** – page 124

Reiki

**The claims that reiki heals through spiritual
energy has no scientific basis, and there is no
evidence that it is anything more than a placebo.**

Like acupuncture, reiki claims to use energies unknown and undetectable to science in its cures, but where acupuncture depends on
the inner human energy of *ch'i*, reiki, which was devised in Japan
in the early years of the 20th century, makes use of something more
like 'the Force' in *Star Wars* – an external universal energy which
is supposed to be channelled by the healer's hands into the body of
the person being treated.

The only positive trials of reiki seem to be those where no controls were imposed – the treatment was not compared with a placebo, and so a natural sense of wellbeing resulted from a belief in
its effectiveness. Although reiki can do no harm, there is always the
danger if it is used instead of a functional treatment in the case of
serious illness that the individual's health will get worse as a result
of not being properly treated.

If you want a placebo-based treatment, over-the-counter homeopathy is a cheaper way to go.

LINKS:
- **Acupuncture** – page 252
- **Homeopathy** – page 295

S

* * * * * * * *

Screens and eyes

For as long as we've had televisions they have been accused of damaging our eyes – but now, with screens everywhere, what's the true picture?

When I was young, older people were always telling us not to sit too close to the TV or to spend too much time watching it. Yet back then, we only had a fraction of the screen time that most of us experience today. You may well spend all day at work looking at a computer screen, then come home to spend the evening in front of the TV. And what were you doing in between these times? Looking at your phone or tablet.

In 2014 there was a scare story that overusing smartphones may damage the eyes. The concern was that the light from LED screens contains more blue/violet light than the natural light that best suits our eyes, and the higher energy blue/violet light can overload the sensors on the retina, increasing the risk of macular degeneration, a common cause of blindness.

The story particularly highlighted the risks from smartphone use, where the screen is held considerably closer to the eyes than a computer or TV screen and is often used in lower lighting conditions, where the dilated irises of our eyes allow more light in.

We have to approach this story with a little caution. There is no study showing that exposure to screens, even smartphones, causes eye problems – this is extrapolating from laboratory-based evidence that light in this region can cause damage to the type of cells used in the retina. So at the very most, this is a precautionary warning as yet.

What is certainly true is that our eyes don't respond well to being focused as closely as a smartphone requires (TVs are usually positioned at a better distance) for long periods of time, and this can cause eye strain. Taking a break from screens for at least five minutes in each hour, and a few longer breaks during the day is highly recommended. Furthermore, regular screen users – which is pretty well all of us – should have regular check-ups with an optician.

LINKS:
• **Eyesight** – page 285

Sunburn

Our whole relationship with the sun is an odd one. We love being out in the sun, but we have always known it could burn us, and now there are worse fears.

There is something very pleasant about sitting out in the sun, and for the last few decades, a suntan has been fashionable. (Before then it was considered much more fashionable in Europe not to be tanned.) We know that sunburn is painful, and that too much sun leaves the skin dehydrated and aged, but the real concern is the rising rate of melanoma – skin cancer.

The problem with too much exposure to the sun is that as well as visible light, the sun's rays contain ultraviolet, a more energetic form of light that can do damage to the contents of your skin cells, triggering cancer. Australia provides a sobering example for the temperate regions of the world. Many Australian citizens, coming from Europe, have moved to a region with two to three times (Southern Australia) or even four times (Northern Australia) the level of ultraviolet in the sunlight. Skin tones balanced for European levels of light fail to give sufficient protection, and the result is an unusually high level of skin cancer.

Even in Europe and North America, climate change may see a considerable increase in the ultraviolet levels we receive. Suntans – and naturally darker skins – are the body's attempt to reduce the amount of ultraviolet that gets through, but their effectiveness is limited. Even the darkest skin pigments are only the equivalent of a factor 8 sunscreen. If we are to spend time in the sun, it is important to use protection.

Furthermore, it seems from a 2014 study that even sunblock does not totally take away the risk of cancer from excessive exposure to the sun. Ideally we should also use hats and loose fitting clothing, or stay in the shade, whenever possible, particularly at the hottest times of day.

Paradoxically, though, we don't want to avoid exposure to the sun *totally*, as sunlight on the skin is our best source of vitamin D. See the **Vitamin D** section for details.

LINKS:

- **Vitamin D** – page 115

Swimming after eating

**It's one of those 'facts' that everyone
knows: swimming after eating is bad
for you. But what's the truth?**

Pretty well anyone will tell you it's not a good idea to swim straight after eating. The advice varies from waiting half an hour to three hours (the long periods come more from Mediterranean countries, where digestion is a more serious business). Exactly what the outcome will be of ignoring this advice varies from stitches and cramps (risking drowning) all the way through to a heart attack. The warning even appears in the original *Scouting for Boys*.

In truth, the whole thing is fictional. There is no evidence of medical problems being caused by swimming straight after eating. Although no formal origin of the story has been found, the suspicion might be that parents on holiday, feeling lazy after a meal, discouraged their children from pestering them for a swim by using this convenient myth, putting it on a par with the fabrication that 'the ice-cream van plays a jingle when it has run out of ice-cream'.

In general it is not a bad thing to avoid heavy exercise for around an hour after eating to avoid the risk of indigestion – but it is not going to put you at risk of drowning.

V

. . . .

Vaccination

Vaccination is one of our best and simplest protections against disease, yet some parents resist it for their children. This is dangerous.

The odd thing about those who resist vaccination is that they are often people who are in favour of 'natural' products and lifestyles. (I am thinking in particular of Hollywood stars who speak out on this kind of subject.) And yet, vaccination is the most natural, and certainly the most effective, treatment against a whole range of dangerous and often life-threatening diseases.

What vaccination does is to kick the body's immune system into action, so that it will protect you from diseases that otherwise would slip past or overpower it. A simplified description of what happens is that either a dead or weakened sample of the bacteria or virus, or a harmless material that also triggers a protective response to the disease, is injected or consumed, and this primes the immune system to attack the disease in question and prevents the illness from ever taking hold.

A great example of the effectiveness of vaccination is the eradication of polio – a terrible childhood disease – from many countries. And vaccination also offers protection against potential killer diseases from measles and TB to the human papilloma virus that is the major source of cervical cancer.

All vaccination carries a small amount of risk of adverse reactions – but these are rare and usually mild – and the risk is far outweighed by the benefits of avoiding these diseases. There is a small but loud anti-vaccination movement, which often spreads messages originated by those attempting to sell some alternative treatment, but that have no basis in medical fact.

It can't be stressed too strongly that the medical benefits of vaccination are very great; there is no evidence for links, for instance, between the MMR vaccine and autism; and that all parents are strongly encouraged to ensure that their children get their full set of vaccinations.

LINKS:
- **Immune system** – page 301
- **MMR** – page 305

Vegetable washing

Because vegetables are healthy food it is easy to think that they don't need the same hygiene standards as other foods – but that can be a fatal mistake.

In 2011 there was a major food poisoning outbreak in Germany, killing 26 people and leaving nearly 3,000 ill. The source was salad – probably beansprouts. It is very easy to assume that food poisoning is always caused by meat like chicken, but almost all fruit, vegetables and salads will have been in contact with soil, which is heaving with bacteria. Always wash fruit and vegetables before eating them, especially if you are going to eat them raw, and wash your hands afterwards.

The recommendation is to rub fruit and vegetables in a bowl of fresh, cold water, then rinse them, rather than just hold them under a running tap, as this can spread bacteria to surfaces. Start with the items carrying least soil. In principle, just as with washing hands, a good scrub with soap and hot water would produce the best clean, but in practice this could contaminate the food, so cold, clean water is recommended. Peeling is also useful to reduce the bacterial load.

One particular myth to be careful of is that organic vegetables don't need washing because they aren't contaminated with pesticide. The level of pesticide residue is not the issue. It is thought that the E. coli in the German outbreak came from products that had been fertilised with animal manure, which is the usual practice on organic farms. Because of this, if anything, organic fruit and vegetables are *more* likely to carry dangerous bacteria; so don't make any distinction between organic and non-organic produce in this regard.

LINKS:

- **Organic food** – page 74
- **Pesticide residues** – page 79

Viruses

One of the most frighteningly efficient sources of destruction in nature, viruses cause some of our most common – and some of our most deadly – diseases.

It's easy to confuse viruses and bacteria. Both are very small and cause a wide range of diseases, but they are entirely distinct. Bacteria are living organisms made up of a single cell that can attack us. Viruses are technically not living at all, but are typically much smaller and simpler structures than bacteria that reproduce by making use of the copying mechanisms in existing cells that they attack. Most viruses attack bacteria, which are much bigger than them, but a range of viruses specialise in attacking animal and human cells.

Because a virus is not truly alive and doesn't have the same structure as a bacterium it can't be killed by an antibiotic, but it can be removed by proper washing and can be destroyed by extremes of temperature. Viruses are responsible for everything from the common cold to AIDS, flu, hepatitis and ebola.

Our main weapons against viruses are to avoid contact with sources, to improve the body's responses to them by vaccination or to use antivirals. These drugs mostly work by providing materials similar to DNA that the virus naturally tends to build into its system in the mechanism that it uses to reproduce, but these special materials are designed to stop the lifecycle of the virus in its tracks.

LINKS:
- **Antibiotics** – page 258
- **Bacteria** – page 261

W

- - - - - - - -

Weight reduction pills

For a long time, the holy grail of dieters was to be able to take a pill that made them lose weight – now we have them. But are they all they seem?

There are a number of weight loss medications or anti-obesity drugs available now, which do what they are claimed to do, but which need to be approached with caution. Apart from being expensive, like all medication they have potential side effects, which, depending on the drug, can range from bowel problems and stomach pains to raised blood pressure. Such products should not be used without medical supervision. Weight loss medication works by either reducing the appetite or stopping some of the food from being absorbed by the stomach, reducing calorie intake.

Probably the biggest issue with weight loss medication (as well as the use of gastric bands and crash dieting), is that it doesn't provide a long-term solution. All these extreme means can be helpful in getting rid of excess weight, but there are only two things that will provide long-term benefit – eating less and exercising more. What is essential is that weight loss products are not seen as a licence to eat as much as you like.

All the evidence is that, psychologically, we find it difficult to treat a quick fix as anything other than a blip in our everyday behaviour, and it can be very easy to put weight back on. It is much better, where possible, to take a gradual, controlled approach that becomes a change of lifestyle.

LINKS:
- **Diet** – page 1

ENVIRONMENT

· ·

Of all the ways that science has an impact on our lives, the environment is the topic where reports in the media can be the most misleading. In areas like health and diet the media has a tendency to respond to a study too soon, without thinking about whether it has any real impact on everyday lives. But when it comes to covering the environment, there is an unrivalled history of journalists making it seem that there is a split of scientific opinion where there is in fact a massive consensus.

So, for instance, whenever the media reports the latest scientific findings on climate change, they tend also to give airtime to someone who disputes the findings (usually without scientific credentials. Those coming up with the counter argument are often politicians). It is as ludicrous as if every time the media reported an around-the-world voyage, they felt it necessary to speak to someone from the Flat Earth Society saying: 'Of course, this couldn't really happen because the world is flat.' At the time of writing, the BBC has been officially cautioned for this practice, because it is misleading – and they are by no means alone.

It is true that there are only some aspects of environmental science where there is a good agreement between the vast majority of scientists. They agree, for instance, that the Earth is undergoing climate change, which has impacts including global warming, increasing numbers of extreme weather events and sea-level rise. While there are a handful of scientists who dispute this, there is no reason for not accepting the view of the vast majority. This is how science works – we go with the best-accepted theory unless new data shows that a different theory matches the evidence better.

There is slightly more variation in agreement over how much of that climate change is man-made – how much is caused, for example, by the increase in carbon dioxide and other greenhouse gases in the atmosphere. There is still a large majority, however, who accept a significant man-made component. There is very little doubt that our industrial and domestic production of greenhouse gases is having an impact on the climate.

Where there is the biggest doubt is over the prediction of how climate change will progress for the rest of the century, and what we should do about it. As the great physicist Niels Bohr said: 'Predictions can be very difficult – especially about the future.' It is impossible to predict the weather more than ten days out. In fact, forecasts beyond this period are *less* accurate than merely saying what the weather tends to be like in a particular location at a particular time of year. Although predicting climate change is more broad brush than a weather forecast, it still involves working with a very complex model of the global environment, running it forward across decades.

The result is that even the best climate models are subject to a lot of uncertainty – which then proves difficult when presenting the results to the public. We know that overall the planet will warm, sea levels will rise and weather events will get more extreme, but it is difficult to say exactly how far things will progress by a particular date.

There is probably even more confusion over what to do about it. Certainly the UK (or any other country) can't do a lot alone. Even if the whole of the UK stopped all carbon emissions tomorrow, the effect on climate change would be almost undetectable. This is why a global response is essential – yet this always gets mired down in politics, resulting in very little action.

Broadly, there are three things we can do in response to climate change (apart from doing nothing). We can reduce our greenhouse gas emissions, so slowing down the rate of climate change; we can try to engineer changes to the environment that will have a counter effect on the climate; or we can accept that we aren't going to stop it and adapt the way we live to cope with a new climate.

In practice, we will do all three to some degree, but there is little agreement over the best balance. It certainly is worth reducing carbon emissions and other greenhouse gas production, but there is a lot of political resistance to the best way to carry on producing sufficient energy with a large reduction in emissions – cutting over to nuclear power – and there is little political agreement over worldwide reductions, particularly from countries like China and India, which are massively increasing emissions as their economies transform, and (probably rightly) don't see why they should suffer more than other nations.

While in principle there are mechanisms to engineer reductions in greenhouse gases, they are fraught with problems. Examples might be to seed the sea with iron, which would increase growth of algae, the biggest natural converter of carbon dioxide to oxygen, or to put vast shades into space that would cut down on sunlight hitting the Earth. But such engineering solutions would have a large impact that couldn't be predicted – the side effects could be worse than the problem being cured – and could also have international ramifications. Imagine, for instance, the outcry if one country put a shield in space that cut out sunlight in another country.

Probably the most ignored solution to date is adaptation. This means accepting that there will be climate change and altering the way we live to cope with it. We have to bear in mind that the climate always has changed and always will be changing, and we have always had to live with those changes if we were to survive. It's worth remembering that we are at the moment in an interglacial – a pause between ice ages – and when the ice has advanced in the past it has wiped out a lot of species and pushed almost all life out of northern lands, forcing them to migrate south.

In a way, man-made climate change has had one positive impact on the future of humanity, as it looks very likely that it will prevent the next advance of the ice. But in return for avoiding the need to cope with that, we have to adapt to deal with increased temperatures and rising sea levels. This means a combination of abandoning some territories, protecting others and using artificial means to moderate the climate.

Of course we do this already. We ought to bear in mind when we see scary forecasts of increases in average global temperature of two to six degrees, that, for instance, Singapore operates perfectly well in a temperature that is over twelve degrees above the global average. The inhabitants do this by adaptation – for instance by mostly living in a city, an environment where air conditioning is more efficient than in a more spread-out population. Adaptation will be hardest where the population is rising and where it is difficult to apply technological solutions. And for adaptation to work, we would need to give serious consideration to a reduction in global population.

However, these are long-term, large-scale considerations. *Science for Life* is all about the impact of science on your life and the lives of your family, and that's what the forthcoming section on the environment will concentrate on.

C

• • • • • • • •

Carbon dioxide emissions

**We often hear about the dangers of carbon
dioxide emissions (or just of 'carbon') –
but what do they involve, and why are
they important for the environment?**

The media, green organisations and even travel companies merrily throw around remarks about carbon dioxide and the environment. We even pay different amounts in the UK to licence a car depending on its carbon emissions. But it isn't always clear what's going on.

Carbon itself is not a problem. All living things contain carbon, you included, and this versatile element also makes up pencil leads, charcoal and diamonds. But when environmentalists talk about carbon, they really mean carbon dioxide. Carbon itself is harmless, but when it is combined with oxygen it produces the gas carbon dioxide. This is produced in relatively small quantities when animals (us included) exhale, but the most significant sources are when a carbon-based fuel is burned. This could be coal, oil, petrol, wood – pretty well all the traditional fuels.

The reason carbon dioxide is a concern is that it is a greenhouse gas, a gas that acts as a kind of insulating blanket in the atmosphere causing the planet to warm. Since the industrial revolution, the amount of carbon dioxide in our atmosphere has shot up, and this is, without doubt, contributing to climate change. We ought to bear in mind, though, that carbon dioxide is not all bad. In fact, it is the most significant food for plants, which extract it from the atmosphere and strip out the carbon to build new cells and exhale oxygen.

LINKS:
- **Carbon neutral** – page 334
- **Carbon offsetting** – page 335
- Greenhouse gases – see **Greenhouse effect** – page 346

Carbon neutral

You will often see companies claiming to be 'carbon neutral' – but what does this mean?

Being carbon neutral simply means that you go through life (and work) without adding extra carbon dioxide, or equivalent greenhouse gases if you are particularly good, to the atmosphere. There are typically three components to achieving this – reducing your carbon emissions, offsetting (where you continue to emit carbon, but pay to have reductions elsewhere) or actively removing carbon from the atmosphere.

There is no doubt that being carbon neutral is better than pumping out carbon at a frantic rate, but it isn't obvious why it is an attractive goal. If you accept that global warming is already quite advanced and will go on getting worse for some time, we don't need neutrality, we need to be in a state where we are reducing carbon levels – all too often, companies use 'we are carbon neutral' as a badge to look good, rather than achieving anything wonderful.

Few of us can get down to zero emissions, so some resort to offsetting, handing off the responsibility for reducing carbon to someone else. But as the **Carbon offsetting** section makes clear, this is easy to get wrong and is never ideal.

By far the best thing to do is to remove carbon from the air, as plants do, but in a more efficient way. This is perfectly possible. One way is 'carbon capture and storage', which removes carbon from the exhaust flues of power stations and traps it in a form that can easily be stored. This isn't something we can do as individuals, but would be relatively simple for large installations – only the technology has had much of its funding withdrawn and hasn't progressed anywhere near as quickly as it should.

LINKS:
- Carbon dioxide – see **Carbon dioxide emissions** – page 333
- **Carbon offsetting** – page 335

Carbon offsetting

Carbon offsetting involves paying someone else to do something about your carbon emissions, rather than cutting back yourself.

The idea of carbon offsetting is simple. Instead of reducing your carbon output to limit greenhouse gases, pump out as much carbon as you like and pay someone else to deal with it. Put like that, it sounds morally dubious – for those with more active consciences, a preferred description is: 'I will do all I can to reduce carbon emissions, but where there is, for example, essential business travel, I will offset that.'

This can be done on the personal level or on a larger scale. When Germany hosted the football World Cup in 2006, they attempted to offset the estimated CO_2 output from travel associated with the games of around 100,000 tonnes by buying carbon credits that resulted in the money being invested in wind power projects in the developing world. Such offsetting has been seen in everything from the 2012 Olympic Games to the flights of British politicians. It has become the fashionable thing to do among the chattering classes. But there is doubt about offsetting's actual value.

There are big issues to consider when deciding if carbon offsetting makes sense. The first is whether or not the action taken provides the benefit that is ascribed to it. The second is how the carbon offset is calculated.

One of the best-known means to offset is to plant trees – but it is also the worst. Though planting trees is a good thing, it is a very slow way to offset carbon, as we have to wait decades before a tree can take a serious amount of carbon out of the atmosphere, but your holiday flight (or whatever you are offsetting) happens right now and will continue to have an impact for years. Worse still, some tree-planting offsetting schemes use existing tree planting programmes. So they take your money, but the trees were going to be planted anyway – in reality, you haven't offset anything.

Other offsetting might use your money to help replace existing dirty power plants with clean generation like wind power. But arguably we need to replace the power plants *and* offset your carbon emissions. The two are not mutually exclusive.

Working out how much to offset is also a problem. There are plenty of online calculators, but they are far too simplistic to give a reasonable measure of what is required, often varying by hundreds of per cent in their recommendations.

At best, offsetting is a sop to our consciences. It is far better to cut your output than to look for a way to offset it.

LINKS:
- **Carbon dioxide emissions** – page 333
- **Carbon neutral** – page 334

Carrier bags

**We increasingly are expected to pay for
carrier bags when shopping – and this makes
sense, unlike some 'green' action on bags.**

Supermarket plastic bags are an easy target for green campaigners. If you don't have a 'bag for life', preferably made out of organic hemp, then you can't be a friend of the environment. And bags for life are great, though it is worth bearing in mind that a lot of people reuse carrier bags as bin liners, dog poo bags and the like. It was interesting that when Ireland enforced charging for bags, the tonnage of plastic film used to make plastic bags consumed in the country went up, as more bin liners, nappy sacks and dog poo bags were sold.

The whole carrier bag campaign is an issue where image is more important than substance. The UK supermarket Somerfield, now part of the Co-op, in an attempt to demonstrate its environmentally friendly approach, came up with the statement that shoppers in the UK use up to 20 billion plastic bags a year, which they said was an average of 323 per household. Unfortunately, this statistic doesn't make sense as it suggests there are more households in the UK than people.

The same statement comments that those 20 billion bags were enough to 'carpet the entire planet every six months.' For 10 billion bags to cover the Earth's surface, each bag would have to stretch across 51,000 square metres. That's over twelve acres. Somerfield also, describing their bags for life, told us that British households 'waste almost six billion plastic bags a year (enough to cover the whole of London).' Leaving aside the inconsistency between 6 billion and 20 billion (perhaps there is a difference between 'waste' and 'use'), someone ought to be able to spot that if 6 billion cover London, 10 billion won't cover the world.

The statistics are entertaining, but there is also very shoddy environmental thinking in the development of degradable carrier bags, pioneered by the Co-op, and supported by UK organic body, the Soil Association. These bags begin to degrade after eighteen months – and are supposed to completely vanish within three years – 'leaving carbon dioxide, water and minerals to be absorbed into the soil naturally.' So where traditional bags lock carbon away for centuries, biodegradable bags give off carbon dioxide and contribute to global warming. An interesting interpretation of being green.

The poor old plastic bag is not only accused of clogging up landfill, but of forming a menace to water-borne life too. Those who have jumped on the bag bandwagon have proclaimed that more than 100,000 marine mammals and sea turtles die each year from getting tangled up in plastic bags or eating them. This simply isn't true. This statistic, nasty though it is, did appear in a Canadian study, but applied to fishing tackle and nets, not plastic bags.

Single-use plastic bags are unsightly and often unnecessary. We should stick to reusable bags as much as possible. But using incorrect statistics helps no one.

LINKS:

- **Carbon neutral** – page 334

E

· · · · · · · ·

Electric cars

**They may have been subject to mockery
from *Top Gear*, but are electric cars the best
environmental solution for the future?**

In an episode of *Top Gear* where the presenters tested an electric car, they made a big thing of running out of battery power and having to charge the car up with a lead running out of someone's window. There are certainly issues with electric cars – and yet if we can solve the problems, they are the best solution to driving in an environmentally acceptable fashion. As long as the electricity comes from low emissions sources like nuclear or renewables, the cars make zero contribution to global warming from journeys, are far cheaper per mile than petrol or diesel, and don't have the distribution network problem of hydrogen cars. So why aren't we all rushing out to buy electric?

We have to accept that for the moment, electric cars are best as runabouts, used for local journeys of less than 100 miles. And that is most journeys for most people. Yes, electric vehicles are useless for a long distance trip, unless battery life can be hugely extended (see below), but they still do those short-term journeys very well.

That being the case, we hit the cost problem. In a radio debate on why the cars hadn't caught on more quickly, industry experts considered how to get over the image of being slow like milk floats or unstylish like a G-Whiz. But the pundits missed the point. Most people know that there are good, stylish, fast electric cars like the Nissan Leaf, the Renault Zoe and the Vauxhall Ampera. But those 'experts' missed the obvious fact that electric cars are too expensive.

Runabout cars – the market electric cars address – typically cost between £6,000 and £12,000 in the UK, but it's difficult to get your hands on an electric runabout car for under £20,000 – and that's with a £5,000 rebate from the government. It doesn't matter how much you save on fuel, it's too much up front. There are schemes to address this by, for instance, renting the batteries (the most expensive bit). But the cars are still hugely overpriced.

Savings come from making cars on a large scale, and arguably car companies need to bite the bullet (perhaps with government help) and subsidise them until sales become sensible. Perhaps they should look at the mobile phone industry, which has made it acceptable to buy expensive phones without noticing how expensive they are by building the purchase into a contract. And obviously far more work needs to be done on making batteries cheaper and longer lasting.

Getting better battery technology, both in terms of higher capacities and in lower use of rare materials, is one of the top ten environmental concerns governments should be investing in – far more valuable, for instance, than building more wind farms. This is something that shouldn't be left purely to commercial ventures. At the very least we ought to see large money prizes – in the millions of pounds – for anyone who can develop a better battery.

LINKS:
- **Hybrid cars** – page 348

F

. . . .

Fire drill

**We all hope never to experience a building
fire, but putting in a little thought in
advance can be hugely beneficial.**

Most of us will never experience anything more than a fire drill, but
should you be in a building when a fire breaks out, a little preparation and organisation makes all the difference.

The starting point is discovering that there is a fire so that you
can take action. Make sure that you have at least one smoke alarm
per floor of your house, and that the batteries are up-to-date. When
the smoke alarm develops that irritating beep, replace the battery;
don't leave it to another day. That will mean you and your family
could go for months without protection. Replace it immediately. An
analysis of fatal fires in London showed that the most common risk
factors were smoking, alcohol consumption, old age and not having
a working smoke alarm fitted.

If no other factors apply, the biggest risk is that at night when
you are asleep, you may wake up disoriented and are more likely to
be on an upper floor. Make sure you know how you can get out of
the house from all the bedrooms. What would you do if the hall or
staircase were impassable? Is it possible to get out of the windows
safely? Plan your exit strategy.

If you are staying in a hotel, take a tip from off-duty aircrew.
They stay in hotels all the time and are trained to make appropriate
preparations. In a hotel you are at a big disadvantage if you need to
get out at night compared with your home, as you don't know the
way around, particularly as you may have been taken to your room
by lift, which you can't use in the event of a fire.

The aircrew routine goes something like this. As soon as you arrive in your hotel room, check the back of the door for routes to emergency exits. Then walk those routes – find your way to the exit for real, as the map on the door doesn't give you a true feel for where to go. Count the doors between your room and the exit. This is because you may have to find your way in the dark, through smoke. If you know the number of doors you can feel your way.

Next time you are staying in a hotel you will probably think 'I can't be bothered' or 'I'll do it later.' This is the importance of undertaking the drill as soon as you arrive. It only takes a minute, but unless done then, you probably won't bother. Hotel fires kill more people annually than plane crashes. Don't begrudge a little preparation that could save your life.

Food miles

**It usually makes sense to buy food that
has been grown locally, as it reduces
the transportation impact on the
environment – but it's not always true.**

If you want to reduce your impact on the environment, it is a good idea to keep an eye on the distance that your food has travelled. With a few exceptions, buying locally grown food will result in fewer carbon dioxide emissions. What's more, local food is less likely to have been stored chilled – it is usually fresher and often tastier. This particularly applies if buying directly from the producer – 'local' food at a supermarket may well have first travelled to a central depot and back to the local store.

However, there are two other factors to bear in mind: miles alone are not enough to make a sensible decision. It's worth finding out if the crop flourishes in your local environment. If it doesn't, then it may be that there is less environmental impact from transport than there is from intensive agricultural mechanisms.

Take tomatoes, for instance. The best way to get fresh tomatoes is to grow them yourself on the windowsill, making use of the fact that your house is heated anyway. But if you buy British greenhouse-grown tomatoes they are less environmentally friendly than Spanish tomatoes, grown without the need for heating. Despite the need for less transport, the British tomatoes result in about three times as much carbon dioxide production as the Spanish, because of that heating. The Spanish tomatoes aren't environmental wonders either – Spanish farms have significantly higher pesticide use than a British greenhouse – but if CO_2 is your main concern and you can't grow your own, go Spanish.

The second factor to bear in mind is that, important though reductions in carbon dioxide emissions are, they might not be your only concern in choosing the source of your food. You might, for instance, want to support African farmers and so be prepared to have their fruit, vegetables and flowers flown over, despite the obvious detrimental effect on the environment. Or for that matter, you might want to support your own country's farmers.

The fact remains that the best way to combine tasty food with low environmental impact is to buy locally grown and raised crops and meat, sticking where possible to the seasonal produce that can be grown with minimal artificial aids.

LINKS:

- Carbon dioxide – see **Carbon dioxide emissions** – page 333
- Greenhouse gases – see **Greenhouse effect** – page 346

Flying

**In environmental terms, there is no good
news about flying – it has the worst effect on
climate change of any means of travel.**

For greenhouse gas emissions, flying is a disaster. Not only does a flight produce a lot of emissions, it does so at an altitude where the gases have a far greater impact on the greenhouse effect.

If we genuinely want to limit carbon emissions, we should give serious consideration to restricting flying to medical emergencies and essential family visits. There is no good reason to fly short-haul – rail is significantly better for the environment, particularly somewhere like France, where electric trains are powered by green nuclear power. It simply isn't practical to build electric airliners in the foreseeable future – planes can never compete. These days, with all the hassle of airports, using the train where possible is also more pleasant and relaxing.

You may think that you deserve your holidays in the sun, and that makes it worth the environmental impact – which is entirely your decision – but you need to be aware of how high an impact than can be. Just one person's share of a single transatlantic journey can result in the equivalent of as much carbon emissions as six months of typical car driving.

Few of us fly more than once or twice a year on holidays – those who really could make the most difference to emissions are those who fly for business. All too many scientists, for instance, (even climate scientists) make regular jaunts to exotic locations for conferences. It would be perfectly possible to run these from local hubs, using videoconferencing – not as good for social interaction, perhaps, but vastly better in environmental impact.

The same goes for the majority of business meetings, which really don't need to be face-to-face. I do most of my business without ever meeting those I work with – it really isn't necessary. We just need to think a little about the environmental impact when setting up an event. For instance, I recently had an article entered for a prize. One of the requirements was that winners had to travel to a US city to pick up their prize in person. To do so is a huge environmental impact for a tiny benefit. It's time they resorted to Skype instead.

LINKS:

- Carbon dioxide – see **Carbon dioxide emissions** – page 333
- Risk of flying – see **Air versus road** – page 214

G

· · · · · · · ·

Greenhouse effect

**The greenhouse effect is caused by gases in the
atmosphere that let sunlight through, but prevent
some of the heat it generates from being re-emitted,
acting as a blanket around the earth. It's not a
bad thing, but it's possible to have too much.**

The greenhouse effect is a mechanism by which some of the heat the
earth receives from the sun is trapped rather than passing back out
into space. It isn't inherently bad. If we had no greenhouse effect,
the Earth would be around 33 degrees colder than it is, with average
temperatures of −18°C, which would be too cold to sustain life as we
know it. We owe our existence to the greenhouse effect. But too high
a concentration of greenhouse gases means the planet can overheat.

When greenhouse gases are present in the atmosphere, most
of the sunlight shoots straight through to warm up the earth. The
earth's surface then gives off lower energy light called infra-red.
Greenhouse gases are particularly good at absorbing infra-red as it
shoots away from the earth, but soon they give it off again in random directions. Some of it heads back towards the earth's surface,
so the gas molecules act like a blanket. (The name is misleading, as
this isn't how a greenhouse works.)

We are used to carbon dioxide being the bad boy of the greenhouse world, but it isn't the only greenhouse gas, nor the most
potent. The biggest greenhouse contribution comes from atmospheric water vapour, while methane – produced by everything from
cow burps to rotting vegetable matter – is 23 times as powerful a
greenhouse gas as carbon dioxide. What's more, there is a real risk
of catastrophic methane emissions into the atmosphere. There is a

huge peat bog in Western Siberia where vast quantities of methane are frozen into permafrost – but as global temperatures rise, the ice is melting, and the bog releases over 100,000 tonnes of methane a day, providing more warming than the man-made carbon emissions of the USA.

Another tricky greenhouse gas is nitrous oxide – laughing gas. Although this has less effect than carbon dioxide, it stays in the atmosphere far longer than CO_2 which tends to sink and dissolve in water. Over a 100-year period after the initial emission, nitrous oxide will have 300 times the warming effect of the same weight of carbon dioxide. The biggest source of nitrous oxide is farming. Plants consume nitrogen, and most fertilisers release some nitrous oxide. The ideal approach is to use plants like clover and peas which 'fix' nitrogen in the soil without fertilisers, but most agriculture needs more than these can provide.

One easy way to reduce fertilisation is to genetically modify plants to require less nitrogen. This is perfectly possible, though it leaves green protestors who don't like both GM and excessive use of fertiliser in a difficult position. This would be well worth doing. If we could reduce fertiliser use by a third it would have the equivalent impact on greenhouse gases of stopping all flying.

LINKS:

- Carbon dioxide – see **Carbon dioxide emissions** – page 333
- **GM foods** – page 61

H

· · · · · · · ·

Hybrid cars

**Hybrid cars like the Toyota Prius were the badges
that movie stars tried to use to demonstrate
their greenness – but how effective are they?**

There was a time when every Hollywood star, trying to promote a green image, would be seen driving around in a Toyota Prius. (Until the cameras went off, when they hopped into their private jets.) There are many more hybrid cars now, sold as a green alternative to traditional cars. However, there are some issues with their green nature.

An average UK driver would reduce their emissions by about 0.8 tonnes of carbon a year by switching to a hybrid. This sounds impressive, but to put it into context, someone driving a shiny new hybrid has a lot of catching up to do. Building an ordinary car results in three to five tonnes of emissions. So simply putting off buying a new car saves a lot more than switching to a hybrid. And building new hybrids produces almost twice the emissions of a conventional car, in part because the batteries are shipped halfway across the world to Japan, then shipped back in the car.

The good news is that a hybrid does have very low emissions – but if you take a look at a comparison with a small diesel, like Volkswagen's Blue Motion Polo, this has even lower emissions than most hybrids.

The other problem with hybrids is that their ability to recharge the batteries when braking is only of value in the stop-start world of town driving. That is the only circumstance when a hybrid is greener than a small ordinary car. If you see a hybrid hurtling down a motorway or twisting around country roads, it is not in its natural habitat. In such circumstances, a hybrid is less fuel-efficient than a BMW 318.

Of course, a hybrid diesel would be better than a hybrid petrol car because diesels are more economical on fuel – but almost all hybrids are petrol. This is because, as Toyota admits: 'Diesel is not a popular fuel in all parts of the world, (especially [the] USA)', and that's why we don't have a better hybrid. Because the American market is not ready for them.

LINKS:
- **Petrol consumption** – page 365

Hydrogen fuel

Hydrogen is often put forward as a green alternative to petrol or diesel for cars – but is it realistic?

Hydrogen cars have had unlikely supporters, from *Top Gear* to Arnold Schwarzenegger during his term as governor of California – but what is all the fuss about, and are they any good? There's good news and bad.

The good news is that, in terms of carbon emissions, hydrogen is a far better fuel than petrol or diesel. When it burns it doesn't give off any carbon dioxide, just water. This has to be managed properly to ensure it doesn't end up as water vapour, which is a greenhouse gas, but assuming that is done, hydrogen puts fossil fuels in their place. But there are some negatives.

First, you don't just dig up hydrogen – it has to be made, typically by splitting water into hydrogen and oxygen. This takes electricity, so using hydrogen is pointless if you use coal to make that electricity. Then hydrogen takes up a lot more space than petrol (which is why we won't see hydrogen powered planes any time soon), restricting space in the vehicle. But most importantly, we don't have a hydrogen distribution network.

It's not easy to get a substance like hydrogen to every filling station in the country. It is extremely flammable – more so than petrol – so filling pumps have to be more secure and more expensive than traditional pumps, and the storage tanks need to be constructed to higher standards than petrol ones. It was hydrogen bursting into flame that ended the career of airships like the Hindenburg. Setting up countrywide filling stations would be a hugely expensive start-up cost.

By comparison, electricity has far more going for it. We already have national and, increasingly, international grids. Electricity is not just available in filling stations; you can fill up at home. Like hydrogen, you need to get electricity from a green source, but that will be increasingly possible. Arguably hydrogen cars are a distraction from making electric vehicles better.

LINKS:
- **Electric cars** – page 339
- **Petrol consumption** – page 365

K

· · · · · · · · ·

Keeping cool

**As a result of climate change, many of us face hotter
summers, with heat levels that we are not used to.
It makes sense to employ science to keep cool.**

During 2003, around 35,000 people died across Europe as a result
of the summer heatwave. And this kind of weather is becoming
more common.

The first essential when it's hot is to get out of direct sunlight,
and to keep the sun off any environment you are in. Blinds, curtains,
shades, sun umbrellas – anything you can use to reduce the direct
impact of sunlight should be deployed. If you don't have air conditioning, your natural inclination may be to fling open the windows,
which is fine as long as it is cooler outside than in. As soon as the
outside temperature is above that of the house, it's better to have the
windows closed. (Investing in an inside/outside thermometer can
be useful, though you can usually feel the difference.)

Be aware of the distinction between air conditioning and a
fan – air con reduces the temperature; a fan doesn't, it simply
pushes the air around. This can result in a comforting cooling
of your skin, as water is evaporated from the skin's surface, but
it does nothing to reduce the temperature of the room. (A fan
slightly increases room temperature, as its motor gives off heat.)
Don't think, by the way, that an open fridge or freezer door will
stand in as a makeshift air conditioner. All a fridge or freezer does
is move heat from inside the box and pump it out the back (the
metal grid on the back is a radiator). Because it isn't 100 per cent
efficient, the device gives out more heat that it removes from the
inside. So running a fridge or freezer with the door open pushes

up the temperature (even though it might feel comfortingly cool in front of the open door).

One of the worst aspects of really hot weather is sleeping. In some of the heatwaves during the last couple of decades in both Northern Europe and the US, the temperature has not dropped below the mid-twenties Celsius mark (mid-seventies Fahrenheit) during the night. This makes it pretty well impossible to sleep. If there's a real heatwave, consider turning your house upside down. The further up the building you go, the hotter it will be, as hot air rises. If you sleep in a ground floor room (or basement), you will be in the coolest place.

You may find that you just don't have room for everything you want to keep cool in the fridge – if so, make use of the cooling effect of evaporation. Get a pair of unglazed terracotta flowerpots, one smaller than the other. Plug up any drain holes and put the smaller pot inside the larger one, filling the gap between with sand. Then pour water into the sand until it is saturated. The water will soak through the porous terracotta and evaporate, lowering the temperature of the pots. Put a water-soaked cloth over the top to complete the DIY fridge. Although this setup will cool most quickly in sunlight, the water will evaporate too rapidly, so keep it in the shade.

Keeping warm

One of the side effects of climate change is more extreme cold weather. It has never been more important to know how to keep warm.

Many of us who live in temperate regions aren't used to really cold weather, and keeping warm is something that we need help to get right. It's obvious to say 'wrap up warm', but what is the best way to do this?

Enter the science of insulation. You are trying to stop your body's internally generated heat from escaping. Multiple layers of clothing provide a great way to do this, as each layer traps some air. This has a double benefit – if you put on your clothes before going out, you will trap warm air in the cavities, keeping your temperature up, and also air makes quite a good insulator. This is why fuzzy fabrics like wool help, because the twists and turns in the fibres trap air.

The advantage of having warm air in the gaps between your layers shows that a traditional warning is an old wives' tale. We are often told not to put on a coat until the moment before we go out or we 'won't feel the benefit'. In fact, you will get better insulation if you keep the coat on, done up for a little while before going out, because the trapped air has a chance to warm up without losing as much to the outside.

Another misleading 'fact' has a much more interesting origin. This is the claim that you lose 50 per cent (or even 75 per cent) of your body heat from your head, making a hat an essential. This was made up for a sales campaign for hats, at a time when it was considered acceptable to make up 'science' to help sales. (This is still done, but it is frowned on more.)

In reality, the most you are likely to lose through your head is about 10 per cent of your body heat. So a hat is useful, but nowhere near as important as many people assume. If you are dressing a baby it becomes more effective, because a baby's proportionately large head can radiate as much as 25 per cent of lost heat.

Try to avoid letting your body come into contact with good conductors, which lead heat away from your body. This is why it is so unpleasant to touch a car on a cold day. In fact, there's the double problem that not only will this help your body heat to leak away, but if it is very cold, the impressive conductivity of the metal will reduce the temperature of the moisture on your fingers very quickly, turning the water solid and freezing your skin painfully on to the metal surface.

The ideal way to avoid this kind of conduction loss is to stay in contact as much as possible with good insulators, like foam polystyrene left over from packaging, or to be in contact with other items that are above the background temperature – another person, for instance. Cuddling up in very cold weather reduces the surface area you have in contact with the cold air. But resist the temptation to down a 'warming' slug of brandy or other alcoholic drinks. Although you feel an inner warming, the actual physiological impact is to encourage heat loss, not reduce it.

L

. . . .

Lightning

**Lightning is a beautiful natural phenomenon,
but it is also phenomenally powerful.
Make sure it doesn't put you at risk.**

We have all experienced thunderstorms at some time in our lives. They are more than a little dramatic. It's not surprising. This is an immensely powerful phenomenon. A typical lightning bolt carries the same amount of power as a power station running for about a second. The temperature near the lightning can get as high as 20,000 or 30,000°C – far hotter than the surface of the sun. As the air shoots up in temperature it expands violently and we hear this shockwave as thunder.

Because thunder and lightning are the same phenomenon, you can use the gap between the two to judge its distance away. The light from the flash travels towards you at 300,000 kilometres per second – practically instantaneously – but the sound moves at a relative crawl, just 340 metres per second. This means that every second you have to wait after seeing the flash is another 340 metres distant. A three second gap puts the storm around one kilometre away, while a ten-kilometre distant flash will take around 29 seconds before it is heard.

Reducing the chances of being hit by lightning is mostly about making yourself less of a target. Get away from high exposed ground, and don't carry anything that makes you taller, like an umbrella. Avoid single trees and metal objects than can act as a conductor. It's natural to go for shelter, but don't stay in the entrance as this doesn't give sufficient protection (and some entrances, notably caves, seem to attract lightning). A car is an excellent shelter as electrical charges can't exist on the inside of a metal box, so it will keep you safe even if there is a strike. Make sure you don't touch the car and the ground at the same time when you get out, though, in case the insulating tyres are keeping a charge on the body.

You can't dodge lightning. Because the air becomes ionised, taking an electrical charge, before the lighting flows, some survivors mention feeling a tingling effect and having their hair stand on end before a strike. If this happens, drop to the ground and keep all parts of your body as low as possible.

Lightning can still cause damage inside a house. Stay away from anything with a wired connection – don't get too near a landline phone, TV or wired internet connected computer or router (wireless connections should be fine).

Low energy light bulbs

In the past, low energy light bulbs got a bad press, but they are getting better all the time, and there is an option to suit every requirement.

No one can sensibly argue it's a good thing to use more electricity than we have to. At the moment, a fair amount of electrical generation in most countries comes from sources like coal and gas, which both use up dwindling fossil fuel supplies and contribute to greenhouse gas emissions. Although electrical power is itself 'clean', the same can't be said for the generation. But even if a country gets all its power from clean sources, we don't want to pay more than we have to – financial savings are the best incentive to go green.

The old incandescent light bulbs, first designed by Joseph Swan in the UK and Thomas Edison in the US used a glowing carbon filament. Since then, the filament has become a metal like tungsten, but there are still two problems – the bulbs burn out quickly and they lose a lot of the electrical energy as heat. Now, admittedly, this contributes to warming your home in winter, but it isn't the best way to provide lighting.

The earlier type of low energy bulbs is not ideal. These 'compact fluorescents' use a refinement of the technology in fluorescent strip lighting. An electrical discharge inside a tube produces high-energy ultraviolet light. This hits a fluorescent coating on the tube, which absorbs the ultraviolet and glows with visible light. Because less of the energy goes to heat, it takes fewer watts of electrical power to produce the same brightness. What's more, a compact fluorescent should last around ten times as long as the equivalent traditional bulb. But there are problems.

The most complained about issue is slowness to reach brightness. The original compacts could take as long as a minute to reach full brightness, which was irritating in some applications (though pleasant in a bedside light). Newer types reach full brightness in around a second, but even so, they lag compared with a traditional bulb. Compacts also have been mis-sold, with exaggerated claims for lifetime and 'equivalent' ratings that compared them with an optimistically bright incandescent bulb. And to add insult to injury, compact fluorescents pollute waste, because they contain a small amount of mercury. Finally, compact fluorescents are bigger than the equivalent traditional bulb and don't fit all light fittings.

In practice, many people tolerate compact fluorescents, but this only seems to be a temporary issue, as they will be as fleeting a technology as floppy disks were on computers. Compact fluorescents are already being replaced by LED lighting, which is better in every respect. The first LEDs were very low power and far too blue for a good room light, but it is now possible to make LEDs powerful enough to light a room and with a 'warm' sun-like colour output.

Where a compact fluorescent typically lasted five times the lifetime of an incandescent bulb (rather than the promised ten times), an LED light can last 30 times as long. It comes on more quickly than an incandescent, and it absolutely slashes the power used. A compact fluorescent might typically take around a third of the power of a traditional bulb, but an LED will take a mere tenth. That's a big saving. At the moment, LED bulbs are fairly expensive, but prices are falling quickly.

N

• • • • • • • •

Nappies (diapers)

**Some people are convinced that disposable
nappies are a huge environmental problem. But
they are convenient. What is the real picture?**

When we had twin babies, I couldn't imagine adding to the load
by using washable nappies instead – and yet that is exactly what
some environmental campaigners urge us to do. You can see why.
Surely it's better for the environment to reuse nappies than to stick
them in landfill? Billions of disposable nappies are thrown away
each year – they take up more landfill volume than any other waste.
But the reality, as is often the case with environmental issues, is not
clear-cut.

The first sign of this came in 1990 when a report was published
that claimed disposable nappies were no worse for the environment
than reusables. There were some issues with the report, which was
sponsored by Procter & Gamble, the company behind *Pampers*,
because it was the kind of study that could be made to show pretty
well anything. In a complex 'lifecycle analysis' like this, there are
so many variables, many of them estimated, that it is easy to influence the results the way you want them to be. To demonstrate this,
a report has been produced that shows that gas-guzzling Hummer
cars are more environmentally friendly than a Toyota Prius.

This was done by stretching a number of the 'guesswork factors' that go into the analysis. The report writers assumed that the
environmental impact of manufacture (high for a hybrid) amounted
to ten times that of the fuel consumed – with no evidence to back
this up. Similarly they assumed that people would keep a Hummer
twice as long as they would a Prius, for a bizarrely specific average

of 34.96 years. Again there is no evidence for the doubling, though high-tech cars are often kept for shorter periods. And the report writers assumed Hummer drivers travelled twice as far as those with a hybrid.

When we look at the Procter & Gamble nappy study, dubious assumptions turn up too. The writers assumed, for instance, that cloth nappy users make use, on average, of 1.9 nappies per change (this is because some use two and others use extra nappies to clean up mess). The study assumed cloth nappies lasted for 90 uses, where cloth manufacturers reckoned they should last 160 times. What's more, the writers assumed disposables were composted – unusual at the time – and that they could assign a cost to the work involved in washing nappies.

A less biased study was undertaken for the UK Environment Agency in 2001 – but again assumptions were manifold. The report makes a big thing about all the measures taken: 'The environmental impact categories assessed were those agreed by the project board: resource depletion; climate change; ozone depletion; human toxicity; acidification; fresh-water aquatic toxicity; terrestrial toxicity; photochemical oxidant formation (low level smog) and nutrification of fresh water (eutrophication).' That's fine, if incomprehensible, but it means there are more variables to guess values for, and doesn't help the analysts assess the relative impact of, say, a disposable being in landfill with a cotton nappy being washed. Each extra category increases the chances of getting it wrong.

The conclusion of the study was that there was no significant difference in environmental impact between disposables and reusables. That's not to say either has no impact – the report suggests that one child, over a period of two and a half years, had impacts (resource depletion, acidification and global warming) that were the equivalent of driving a car between 1,300–2,200 miles. There are no green babies.

So, there is no good scientific evidence to help make the choice, and the best guidance we have is there isn't much difference. It's really down to personal preference.

LINKS:
- **Hybrid cars** – page 348

Natural radiation

We associate radiation with nuclear power stations and atomic bombs, or phone masts and Wi-Fi. But we are exposed to radiation from our environment all the time.

Radiation falls into two categories, which are separate, though often confused. *Electromagnetic* radiation is light, both the visible kind and the light that goes beyond the visible spectrum and includes radio, microwaves, ultraviolet, X-rays and gamma rays. There is also radiation consisting of high-energy *particles*, typically produced by nuclear reactions. Both the most powerful electromagnetic radiation, like X-rays and gamma rays, and particle radiation are bad for us. We are rightly concerned about this when there is a leak from a nuclear power station, but such radiation is present at all times.

In normal areas, the background levels of radiation appear to do us no harm. (Even if they did, we couldn't avoid all radiation whatever precautions we took. Our bodies are slightly radioactive, so radiation from one part of the body is always impacting the rest.) But when levels rise, radiation increases the risk of some cancers and at high levels can cause illness and death. Such radiation contains three components – alpha particles, which are big, relatively slow particles and are stopped by the skin; beta particles, which are faster electrons; and gamma rays, which are extremely high-energy X-rays and which form the electromagnetic radiation produced by the nuclear processes.

There are a number of ways that natural radiation levels can be higher than normal. Every country has varying levels depending on the rock formation. In the UK, for instance, Cornwall has three times the background level of London, and the granite rock (typical in areas with higher background levels as it contains more uranium) gives off radon gas, which can collect in a house and push up radiation to dangerous levels; so it is important that houses there are suitably ventilated. If you are concerned, test kits are available from www.ukradon.org, or check your local council for advice.

The UK has low background levels. Such radiation is measured in milliSieverts (mSv) per year. The average background level in the UK is 2.5 mSv/year, rising to around 7.6 mSv in Cornwall. Other countries have higher levels. France averages 5 mSv, the US 6.2 mSv and Finland over 7. The heavily populated area with the highest background radiation is in parts of Kerala and Madras in India, where levels are around 30 mSv per year, while a few remote regions can reach as high as 50 mSv. The general recommendation is that levels are kept below 20 mSv per year.

X-rays (and X-ray scans like a CT scan) also add to your dose, which is why they are used sparingly. Perhaps more surprisingly, eating shellfish and flying both increase your exposure. Shellfish do so because they accumulate heavier metals, including radioactive material. They will add a typical 0.5 mSv to the annual dose of a regular shellfish eater. As for flying, this is because natural radiation also comes from the sky. Some originates from the sun and some from deep space particles known as 'cosmic rays'. When you are in a plane, there is less air between you and this incoming radiation, providing lower levels of protection.

To get a feel for the impact, a single flight across the Atlantic gives you the equivalent to having around three chest X-rays – nothing to worry about as an occasional event, when you consider that a chest X-ray weighs in at around 0.02 mSv, but if you commute by air on a daily basis (or fly as a job), there is a small but significant increased risk.

The level of radiation in the air can be strongly influenced by the solar cycle. The sun goes through phases of putting out more or less radiation in a roughly eleven-year cycle. At the peak of a solar storm, the levels of radiation can be 100 times the typical levels – very frequent flyers might like to consider reducing their flying hours during solar peaks.

Nuclear power

There are a lot of confusing messages about nuclear power, which often ignore the science.

For a long time we have had confused messages from governments and campaign groups about nuclear power. It's not surprising when we are talking about nuclear fusion, the power source of all current nuclear power stations, as this is related to the mechanism of an atomic bomb – though it is worth stressing straight away that it is physically impossible for a nuclear power station to blow up like a nuclear bomb, which uses a different form of uranium to the reactor.

There are certainly some issues with nuclear power. It is expensive to set up, though it would have been significantly cheaper if we had ever taken nuclear power stations into mass production. And nuclear waste has to be disposed of carefully and kept safe for a very long time. Finally, there is the risk of 'another Chernobyl'.

Chernobyl in the Ukraine was by far the worst nuclear accident we have ever had, caused by the combination of a badly designed reactor and a ludicrously poor safety regime. Yet even so, coal kills far more people each year than have died as a result of Chernobyl, through mining deaths, secondary deaths (as in the Aberfan disaster in Wales), and through lung conditions from pollution. The reality is that the fear of nuclear accidents and the media coverage they get means we give them far more weight than we should.

It is notable that the reactor problems at Fukushima in Japan got more media attention than the tsunami that caused them. The reactor leak killed no one, but the tsunami killed thousands. Yet in a panic response to the publicity, Germany withdrew from its nuclear programme, significantly increasing its carbon emissions.

On the positive side, nuclear energy is green and low in carbon emissions. If the projections of the impact of climate change are anything to go by, the deaths likely to be caused by this (the last heatwave in Europe killed 35,000 people directly and another 15,000 from pollution) make climate change a far greater issue than nuclear accidents, especially as much safer nuclear technology than was used in the old reactors now exists.

The other essential is that we don't let our view of nuclear fission colour our impression of nuclear *fusion*. Fusion is the power source of the sun, and work is underway to bring the fusion process to a stage where it can be used to generate power. This is intensely difficult, but current estimates are that we could be generating energy this way on a large scale by 2050. Fusion does not produce high-level waste and uses readily available fuel. Given that solar, wind and wave can only do so much towards giving us energy security, fusion is our best hope for energy sources in the future.

LINKS:

- **Carbon dioxide emissions** – page 333

P

• • • • • • • •

Petrol consumption

**Motoring fuel costs are set to continue rising.
Using a little science to minimise usage will
help your budget as well as the environment.**

You don't have to understand how a car works to be aware of the
basic science behind fuel consumption. The more energy that is
used, the more fuel you have to buy – and that energy goes on
moving the car, fighting against wind resistance and powering the
goodies in your vehicle.

A starting point is to give more consideration to fuel consumption when choosing your next car. Cars are now available with
impressively high mileage. If you only do short local journeys, consider an electric car – prices, though still high, are coming down,
and the cost per mile is a fraction of that for a conventional vehicle.
Otherwise, a high mpg diesel is often the best choice – certainly
better than a hybrid for many uses.

Then try to keep energy use down. The faster you go, the more
energy it takes. Driving at 56 miles per hour uses 25 per cent less fuel
to cover the same distance as 70. Similarly, on the motorway, you
use 30 per cent less fuel at 70 than you would at 85. Heavy-handed
acceleration and braking also increase the usage. You can lower fuel
consumption by up to 30 per cent simply by driving as smoothly
as possible.

Don't hang around in a low gear. At 37 miles per hour, a car uses
around 25 per cent more fuel in third gear than in fifth. Your tyres
can also help. There are now tyres specifically designed to reduce
fuel consumption, and any tyres will increase consumption if they
are not inflated to the correct pressure. One temptation is to slip

the car into neutral when going downhill. This uses more fuel than staying in gear with a fuel injection car (that's almost all modern cars). However, you can save a little by turning the engine off if you will be stopped for more than 15 seconds, say at traffic lights.

Air resistance increases fuel consumption massively. Take any bike racks and carriers off the roof and drive with your windows closed. Although you shouldn't use air conditioning unless you have to (it uses a fair amount of power), it is more fuel efficient to drive with air con on than it is to turn the air con off and open the windows.

It might seem obvious, but don't carry more weight than you have to. Check what's in your boot, for example, and if your car has removable seats, make sure you really need all of them all the time. It is also worth filling the tank to the half way mark, rather than all the way up, if practical, as otherwise you are hauling around quite a weight of fuel.

The absolute best way to reduce fuel consumption is not to use the car. Can you walk instead? All of us are guilty of making unnecessary short journeys, and there's the added bonus that cars consume up to 60 per cent more fuel per mile in the first couple of miles before they are warmed up, so cutting out short journeys is particularly beneficial. For longer range, check out a bike or public transport. Of course, a car is often the only sensible solution, but don't assume it's the only way to get around.

LINKS:

- **Electric cars** – page 339
- **Hybrid cars** – page 348

Phone masts

Phone masts have been blamed for a host of ill effects, but there is no evidence of them harming humans.

Like Wi-Fi, phone masts are blamed for a multitude of ills. Also like Wi-Fi, a phone mast is a radio transmitter, in this case for the radio signals used to communicate with mobile phones. A typical phone mast puts out considerably more power than a Wi-Fi hub, but it will also be sited at a greater distance from passers-by.

In studies on phone mast problems, nearby residents have reported 'significantly higher occurrences of headaches, memory changes, dizziness, tremors, depressive symptoms and sleep disturbances than a control group'. However, these are typical symptoms that are generated by the nocebo effect, the negative version of the placebo effect.

A common factor in the studies that have shown a connection between phone masts and these symptoms is that the studies are poorly controlled. They often involve very small numbers of self-selecting people who are already reporting problems with phone masts. Also, there is no attempt to control the data, for example comparing the symptoms of those who have an inactive phone mast sited near their house. Where proper controls are imposed, there is no difference in impact between technology that is not functioning and active masts.

Despite many studies, there is no good scientific data linking phone masts and any kind of illness. One problem here is the cluster effect. Randomly occurring events, including cancers, occur in clusters, not evenly spread out. Inevitably some of these clusters will be near phone masts, and it is easy then to jump to the incorrect conclusion that the phone masts caused the events.

LINKS:
- **Clusters** – page 221
- **Wi-Fi** – page 376

Power cuts – computers

Most of us face power cuts occasionally, particularly in rural locations. Nothing likes a power cut less than your computer.

We have become used to an electronic, connected world – a power-cut quickly takes away all that connectivity and leaves us painfully isolated. But worse, it can do damage to your computing equipment.

While laptops, tablets and smartphones will be okay, at least as long as their batteries last, desktop computers will not only stop working immediately, but may lose whatever you were working on. And even if your portable equipment is fully charged, it has limited use if your Wi-Fi has lost power.

If your technology is important to you – especially if your work depends on it – it is worth protecting it with an uninterruptible power supply (or UPS). This keeps the power going for a short time (typically fifteen minutes) after a power cut. It does this by keeping a battery charged up and switching over to that to keep things running. On a desktop computer this means that you can shut down programs properly rather than having the plug pulled mid-sentence, while it also keeps your connection to the internet live for a precious few minutes to find out what's going on. This assumes that your internet provider has not also lost power, but most providers have independent power supplies.

Even if you only use battery-powered devices like a laptop or tablet, there is the potential for problems if they are plugged in to a charger when the power comes back on as there is often a spike – a surge in voltage – that can damage electronic equipment. The same problem often arises during a thunderstorm. The ideal is to have your equipment plugged into a surge protector. Most UPS devices have these built in, but you can also get cheaper extension cables with surge protectors, and they are strongly recommended to keep your precious electronics safe.

LINKS:
- **Lightning** – page 355

Power cuts – fridges

A power cut is inconvenient, but once it lasts more than a few hours it puts your chilled and frozen food at risk. Science can help keep things cool.

As soon as a fridge or freezer is without power it's down to the physics of heat transfer to keep your food safe. Modern appliances are well insulated, so the first essential is not to open the door more than you have to. Every time you do so, you let a blast of warm air in and lose some of that precious chill. A modern freezer that is well filled (frozen food keeps cold much longer than air) can stay frozen for up to 48 hours if the door isn't opened. Half full, it won't last longer than 24 hours.

If your freezer hasn't much in it, it's worth transferring any ice, cool bag blocks and unwanted frozen food into the fridge to help keep the food in there chilled. Either way, when you have to open the fridge or freezer, if you have any blocks of foam polystyrene about the house (often found in the packaging for electronics, etc. – you may have some in boxes in the loft), use them to minimise the air space, which will reduce the heat getting to the chilled food.

If you have a food thermometer, check the temperature of the food when the power comes back on. Frozen food should still have ice crystals in it and should not be above 4°C. If that's the case it should be safe to refreeze (though the texture of some foods may suffer). Any perishables in the fridge that have gone over 4°C for more than two hours should be thrown away. It would be better, though, if you have some means of cooking without power, to cook food before it reaches that limit, rather than waste it.

Not everything you have in a fridge is perishable. We have a tendency to keep food in the fridge these days like vegetables, eggs, preserves, pickles and sauces that are perfectly happy for some time at room temperature. There should be no need to discard these and other room temperature comfortable contents.

If power cuts are long-term (at the time of writing, thousands of houses in the UK were without power for at least a week as a result of storm damage), look for alternatives to keep milk and perishable food cool. If it's cold enough, your garage or shed may be at a temperature comparable to the typical 3–4°C of a fridge – consider using this, but you will have to keep an eye on the temperature. If conditions are warm, it might be worth constructing a few flowerpot fridges (see **Keeping cool**).

LINKS:
- **Keeping cool** – page 351

Power cuts – lighting

**Power cuts have a nasty habit of coming at night,
which is where we quickly discover that lighting is
one of the most essential uses we make of electricity.**

Especially if you are home alone, a power cut at night quickly demonstrates that lighting isn't just about being able to carry on doing things after dusk; it also gives us a sense of security.

The first line of defence is torches (flashlights). Make sure that you have spare batteries and ideally at least one wind up torch. It's also worth thinking about just how you can get to your torch in the dark. Make sure that there are one or more torches in easily accessible locations.

Modern LED-based camping lanterns are ideal to back up those torches for a longer power cut. If your area suffers from regular power cuts, you might also consider getting one or more large lead acid batteries, which will run a light for a good long period of time. Be careful to get the right type. Lead acid batteries come in two varieties. Car batteries are designed to give a very large current for a short period of time, and are intended to be kept topped up most of the time. By comparison, caravan batteries (known as deep cycle batteries) can stand being run down then recharged many times: these are the type you need.

To keep a lead acid battery charged, consider a solar charger, which has the advantage of working even if there is a power cut. Be careful, by the way, with these batteries. The acid inside is highly corrosive. Also the batteries can give off dangerously flammable hydrogen when being charged, so keep them in a well-ventilated space.

The traditional alternative to electric lighting is of course candles. Families managed well enough with these and oil lamps for hundreds of years, but there were a large number of resultant house fires – and we aren't so used to candles these days, so you need to be extremely wary. Make sure candles are carried very carefully, and that any candles in bedrooms are positioned securely and not left alight when your family go to sleep. (Similarly, don't leave them alight if you leave the house.)

Never use a candle, even a tea light, as a night light – it is simply too risky.

If you do use a candle, don't waste what little light it gives off – put a reflector like a mirror behind it.

S

. . . .

Spiders

**Most countries outside Australia have
harmless spiders – and yet most of us are
made uncomfortable by them. Why?**

Around 10 per cent of the population are arachnophobes – people
with an extreme, irrational fear of spiders – but far more of us have
a degree of irrational fear (assuming we are in Europe or most of
the US; Australians, for instance, have a very reasonable fear). After
all, there are no dangerous European spiders, and hardly any bite
humans. So why does the spider in the bath give us such a scare?

It seems to be due to the separation of roles in different parts
of the brain. The information from the eyes is passed to two parts
of the brain, one of which handles the thinking role, the part that
enables you to decide rationally that a spider is really no threat to
you. The other receiving part, though, is an almond-shaped chunk
known as the amygdala (there are two of these, one each side),
which has a wide range of roles involved in storing emotionally-
based information and sorting out the memory. Another job of
the amygdala is in the 'flight or fight' response, where it can send
high-speed instructions to take action (such as making a rapid exit
from the bathroom).

The connections that trigger this response are primarily unconscious. They will literally kick in before you can think about what
you are responding to – not a bad thing when dealing with life-
threatening danger. It's not totally clear why we are scared of spiders
in the first place, but it may well be because humans originated in
an environment where there were dangerous spiders and we haven't
had time to evolve away from this 'pre-programmed' reaction.

Incidentally, while we're dealing with the science of spiders, it's a myth that they come up the plughole. Most houses have a fair number of spiders here and there, and every now and then one will wander on to the edge of the bath. They find the slippery surface impossible to get a grip on, so take a ski ride down into the bath and can't get out again.

Street violence and architecture

Most of us are occasionally wary out on the street. But what makes a neighbourhood particularly dangerous?

It might seem surprising, but the architecture of where you live can influence the risk of being attacked in the street. A study has found that the streets around a high-rise block of flats are more dangerous than those around traditional low-rise accommodation.

This has nothing to do with the relative poverty of the inhabitants or other factors, which were accounted for in the study. Instead, it seems to be due to the nature of tall buildings.

Although there are occasional exceptions where the attackers are looking for publicity (terrorists spring to mind), most attackers, whether muggers or worse, prefer quiet streets where people are unlikely to spring to the assistance of the attacked. We generally feel safer in a crowd, because the others around us are unlikely to stand by while we are assaulted (we hope).

The increase in risk seems to be because with a surrounding of low-rise buildings there are likely to be more people who are looking at the street, even if they are in buildings – where a high-rise puts most nearby people out of rescue range. The study even threw up an approximate way to calculate the risk. Each extra floor meant another 2.5 per cent chance of being mugged or having your car stolen. So a twelve-storey block of flats meant you had an extra 25 per cent chance of being the victim of crime than a traditional two-storey building. Just feel sorry for those who live in the shadow of the Shard, the Empire State Building or the Burj Khalifa.

Most of us would agree that low-rise living makes for more pleasant, human scale spaces – but it seems it even makes the neighbourhood safer.

W

· · · · · · · ·

Wi-Fi

**A wireless network (or Wi-Fi) incorporates
a low power radio transmitter, which has
caused concern to some householders,
but there is no risk from using them.**

There are regular scare stories in the media about people suffering from side effects of using a wireless network (Wi-Fi) system. A wireless router – now found in many homes and public sites like cafés and workplaces – converts internet data into radio waves, which are broadcast from a very low power transmitter. Just as is the case with phone masts and mobile phones, it is common for anti-Wi-Fi propaganda to use the loaded term 'radiation' when referring to the output of a wireless router. This is true, but misleading, as the usual association of 'radiation' is with the potentially hazardous nuclear radiation.

The radiation from Wi-Fi is electromagnetic radiation, which is another term for light. Light comes in a range of frequencies, from radio, the lowest frequency, through to microwaves, visible light, ultraviolet, X-rays and gamma rays. Usually, the higher the frequency, the more potential for damage. (Microwaves are a special case, which despite being relatively low frequency, happen to coincide with the vibrational frequency of water and so cook food well.) We have radio signals travelling through us all the time from TV, radio and mobile phones without any concern. Wi-Fi merely adds to this mix.

You will sometimes see studies quoted as evidence that there are health risks from Wi-Fi, but these studies are poor quality and are usually based on participants telling the researcher how they feel, rather than any qualitative or controlled attempt to detect any impact from the presence of Wi-Fi. (This research is often actually based on phone masts, but used as evidence against Wi-Fi.)

Some individuals claim to suffer from 'electrosensitivity', getting headaches and other symptoms when a Wi-Fi network is in use. This can be tested with proper controls, asking the individual to tell whether the Wi-Fi is switched on without them or the experimenters knowing if this is the case, and thus providing the 'double blinding' needed to make an experiment like this useful. This has now been done in a range of trials and the evidence is clear that electrosensitivity is imaginary. Those suffering from it may well *believe* that they can sense the presence of Wi-Fi, but in reality their success at guessing when Wi-Fi is operating is no better than random chance.

LINKS:

- **Phone masts** – page 367
- **Natural radiation** – page 361

FUN

• •

This short section wasn't in the first draft of *Science for Life*. But on reading it through, I got the sense that the book seemed to be saying that the benefit of science to life is all about serious things. Okay, we're thinking about things like health and exercise – good things, certainly – but we are often rather earnest in our attempt to get things right. It only seemed reasonable that we should also address the matter of fun.

I can see readers cringing instantly, thinking that I am going to suggest a scientific formula for fun, which sounds as likely as those apparently scientific formulae you see in the newspapers for the happiest day of the year, or for who has the sexiest bum. I'm sure you realise it already, but these formulae have nothing to do with science. They are the products of PR companies, looking to get publicity for a sponsoring company on a slow news day. Often the 'formula' is dreamed up by the PR company, which then hunts around for a scientist they can pay to justify it, and scientists, being human (and underpaid), will sometimes go along with it.

Relax, though. I am not going to produce the scientific formulation to generate fun. What I mean rather is to look at areas where science can help with fun activities, whether by making them cheaper or more approachable, or just giving us a greater appreciation of the mind-boggling stuff going on behind the scenes that we take for granted.

There will also be one or two entries that we hardly think of as 'fun' in the general way of things – queuing springs to mind. Here, what I mean is more that this is something where we can get some fun from thinking about the science involved in a mundane, everyday activity.

Buses come in threes

We've all been there. You wait for a bus for ages, then several come in the space of a couple of minutes. Is it true, or are we fooling ourselves?

It's easy enough to fool ourselves when it comes to statistics. For instance, we all have bad days when everything seems to go wrong. But it's not so much a curse on that particular day, it's just that once we start to suspect the day of being a bad one, we notice every little thing that goes wrong, even though it might happen on another day and be quickly forgotten. We become sensitised to the particular occurrence, and see more of them. (It's similar to the way that once we notice a few white cars, there suddenly seem to be loads of white cars on the road.)

Even with buses there will be a bit of a sensitisation effect. We expect buses to come in clumps, because it's said so often that it has become a joke, and so we are probably more aware of this happening than we would otherwise be. But there is a good reason why you might expect this to be true.

Buses are scheduled to run at regular intervals. But a bus that has lots of people getting on and off will start to fall behind, where one that has few passengers will stick to the schedule. If you imagine the first bus that turns up after quite a long wait, there will be lots of passengers waiting at the stops. So that bus is slowed down by picking people up. This means that the gap closes with the next bus behind it. As the next bus will now arrive soon after the first bus, it finds fairly empty bus stops. So it continues to catch up. It's a self-fulfilling prophesy. Buses really do naturally travel in packs.

Crop circles

Enthusiasts consider it strange that science won't take crop circles seriously – but what is the story behind these strange field markings?

Crop circles are markings in crop fields, created by pressing down the crops to produce sometimes intricate and beautiful patterns that can only be truly appreciated from the air.

In the early years of these phenomena, which started 'cropping up' in the West of England in the 1970s, it was speculated that they could be produced by strange weather patterns like freak whirlwinds, or even that they were messages left by alien visitors. Many agreed that the designs couldn't have been made by human beings without causing more damage to the surrounding crops. But in 1991, Doug Bower and Dave Chorley admitted that they had been responsible for starting the crop circle craze.

They demonstrated how the circles were constructed, by flattening portions of a crop using one or more planks of wood with ropes attached to the ends. To make the apparently precise alignments, an old hat with a wire loop attached was used to take bearings on landmarks. And because they used no heavy machinery – just a couple of men on foot – there was no disruption to the surrounding crops. With constructions this size, the eye is reasonably forgiving, so the precision often appears better than it actually is.

Many others took up the crop circle gauntlet, producing more and more complex designs. Geometric patterns are still the most common, though circles have also been made in the shape of advertising logos. Despite this, a number of believers cling on to the possibility that *some* crop circles have extraterrestrial origins. In principle this could be true, just as, in principle, the supermarket at the end of the street could have been built by aliens if you never saw it being built by people – but it's very unlikely.

The scientific view of crop circles is simple – they are made by human beings for fun. However, the disappointing origins of crop circles shouldn't take away from their excellence as pieces of short-lived art.

Digital media

Many of us have increasing amounts of our media in a digital format – but have you ever thought about who it belongs to?

The chances are increasingly high that you will own digital music to play on a computer or a phone, ebooks and digital copies of films and TV programmes, rather than CDs, DVDs and paper books. But digital media come with a challenge. What happens if you want to change their ownership? You can give someone a book or a CD, but what do you do with your digital collection?

There was a minor media storm a while ago over whether you can leave your digital collection in your will, allegedly because Bruce Willis wanted to make sure he could leave his iTunes collection to his children (this proved to be a hoax). And it is true that there is an issue with what will happen when the holder of digital media dies. But many of us face more immediate problems. If, for example, a couple splits up, traditionally there would be the ritual of going through the books and records saying which were whose. Now, though, there is a need for a mechanism for making that split in a joint digital collection.

Similarly, a lot of us will have set up our children on a shared account, as they don't have the credit or debit card necessary to pay for downloaded music or ebooks. But again, what happens when your young adults leave home?

There isn't a satisfactory reply. Although it would be perfectly possible for the big players like Amazon and Apple to provide a mechanism to move a track or ebook to someone else's account, this doesn't currently exist, and the companies regularly refuse to comment on it. Their licensing agreements make it clear that you can't pass on the licence. Someone else can take over your physical device – so, for instance, you could give someone a phone with your music on – and control of an account could be taken over by someone else. But there is no right or mechanism to transfer tracks or ebooks from one person's account to another's.

It seems likely that in some cases the only option will be to start again from scratch. However, if you are setting up children on something like iTunes, it would be worth ensuring they have their own account funded by your card (you don't have to give them the password) rather than sharing an account.

Hopefully, by the time most of us are coming to the stage of wanting to bequeath our digital collections in a will, there will be a legal requirement for the companies to make this possible … but don't hold your breath.

Falling toast

It's one of those popular statements about life that toast always falls butter side down – but unlike many of them, this one has a scientific basis.

We've all heard of Murphy's Law, or Sod's Law, that if something can go wrong, it will. And we have plenty of opportunities to spot this in action – but often, we are fooling ourselves. This is because memory is selective. We only remember things that are especially good or bad or surprising. (This is why 'spooky' coincidences happen. We don't remember all the times we think of someone and they don't ring two minutes later, only the very occasional times we think of them and they do.)

However, there is one instance where real science comes into play, and that's the whole business about toast falling butter side down. Admittedly, there was an 'experiment' that seemed to show this wasn't true, but the giveaway was that this was not a proper scientific study, but an exercise done for the BBC TV show *Q.E.D.* in 1991 (or possibly 1983 or 1993 – sources disagree). The experiment showed that toast fell butter side down 50 per cent of the time – as you would expect when tossing a coin. But unfortunately, like a lot of TV 'science experiments' the demonstration was rigged to show the answer the producers required.

Their aim was to disprove Murphy's Law – but the problem is that the butter side down effect is not a myth, it's basic physics. What the BBC did wrong was to treat the slice of toast as if it were a coin, tossing it high in the air. When we toss a coin, with a quick flick from the thumb, it rotates many times as it flies up in the air before eventually coming down to the ground. That's enough of a randomising effect to get an approximate 50:50 result. And the same went for the BBC's toast.

However, what happens in the real world is that toast slips off your plate, most often when you are picking it up. At this point the toast is around waist height, and as it slips over the edge of the plate it begins to rotate. However, it is rotating relatively slowly and doesn't have far to fall. As a result, the usual outcome is that it only has a chance to make a half rotation (give or take a quarter rotation either way). So, because it started butter side up (unless you have a very strange way of putting toast on your plate), it ends up butter side down. Q.E.D., as the BBC might say.

Ghosts

**In the past, a number of eminent scientists were
interested in ghosts and the whole paraphernalia of
haunting. But what's the current scientific view?**

Ghost hunting grew strongly in the early part of the 20th century, fuelled in part by large-scale bereavement caused by the First World War. Over the years, there has been an increasing use of scientific equipment on ghost hunts, which makes it look as if this were a scientific study. But most ghost hunts have no science involved.

The scientific consensus is that ghosts do not exist. There is very little good evidence for them, and the various meters and detectors used by 'ghost hunters' are ones that respond to perfectly natural changes in the environment that require no ghosts to make them happen. Of course, there is plenty of anecdotal evidence of ghosts and hauntings, but these can usually be explained fairly simply. A combination of imagination, hallucination and fakery works as a much more likely explanation that actual haunting. Good, well-captured evidence is hard to come by, and until there is any better evidence, this will remain the scientific consensus.

You might wonder, then, how it is possible for programmes like *Most Haunted* on the TV to show clear evidence of ghostly goings-on? These are entertainment shows, not scientific investigations. There is well-recorded evidence of presenters faking various items on the show, a classic being a scene in which a table begins to move of its own accord during a séance. This is easily found online and worth watching.

In the video, the table begins to rock from side to side, vibrating with uncanny regularity. It is as if some unseen force is driving it. The large candlestick in the centre of the table vibrates as the tension rises in the voices of those taking part. Clear evidence? Not if you look closely. Take a look at the tablecloth and you will see that it is rippling rhythmically under presenter Yvette Fielding's fingers. Exactly as if she were pushing the table to produce the effect and the pressure of her fingers was crumpling the cloth. Other clips catch presenters pushing objects with their feet or simply reacting to something that isn't there.

Ghosts are fun and fascinating to read about or watch shows about on the TV – but shouldn't be taken too seriously.

Hyaluronic acid and other wonders of chemistry

There's nothing cosmetic companies love more than throwing in 'the science bit' – yet we should take it all with a huge pinch of salt.

You may be surprised to find cosmetics under the 'fun' section – but it's hardly about health or diet, and in the end the message of examining the science of cosmetics is that it is primarily about enjoyment. If you enjoy using a cosmetic, fine. But beyond the obvious benefits of a basic moisturiser, don't expect any truly scientifically verifiable improvements from slapping on expensive potions.

Most of the science bits in cosmetic advertising rely on taking a word that sounds suitably imposing (DNA comes up a lot) and splashing it around liberally in the hope that it makes the product sound better. DNA, for instance, is in every living thing, so if that's helpful for beauty improvement, you might as well smear banana on your face.

There didn't seem a lot of point going through all the verbal armoury of the cosmetic world, partly because it changes every week and partly because the answer is pretty much the same for everything. But I would like to pick out one of the most interesting (and frequently used) ingredients of expensive cosmetics, hyaluronic acid.

This is a naturally occurring gunk that was first found in the vitreous humour, the jelly-like substance inside your eyeballs. ('*Hyalos*' is Greek for vitreous. Thankfully the hyaluronic acid in cosmetics is artificial, rather than eyeball gunk.) It was found to occur widely in the body, where it forms part of the matrix, the structure that helps support the cells in the body.

The substance has some medical applications, to help with tissue healing after specific operations – and its first cosmetic application was as a filler, injected to expand the soft tissues and smooth out wrinkles. (This is a cosmetic treatment that can be risky if carried out by anyone other than a medical professional.) But these days, the acid crops up in many skin products.

There are two reasons for this – one that works and one that seems doubtful. The realistic reason is that hyaluronic acid is a 'humectant', which means it traps water vapour from the air, making it a good moisturiser (though there are plenty of cheaper moisturising products that are just as effective, like simple Vaseline). The main reason the substance is given such prominence, though, is to leap from the way it supports cells in the body to suggest that it will 'stimulate collagen synthesis' and plump up the skin, reducing wrinkles. This isn't supported by clinical evidence. There is also a concern with applying it to the skin that it can break down under strong sunlight to form irritants, though I'm sure the cosmetic companies will have this covered (probably by putting so little hyaluronic acid in the product that it doesn't matter).

One last use is highly doubtful to have any benefit. You can buy hyaluronic acid food supplements. These don't seem to have value as the acid will break down in the stomach and won't survive to be somehow miraculously transported to the skin. More worryingly, there are possible adverse effects for those who are pregnant or breastfeeding, so it isn't clear why this substance is available as a dietary supplement.

Irritatingly itchy

You may have wondered why it is that you tend to get an itch at certain times – and science has an answer.

Have you ever wondered why it is that every time you hear there has been an outbreak of fleas or head lice at a school you suddenly feel an irresistible itch? And for that matter why it is that you often get an itch when you are trying to be still, for instance when you are listening to a concert of serious music that hasn't really got your attention? Or when you've been told to be really still while they take an X-ray?

There are two things happening here, both connected to your brain, rather than the nose or arm or scalp – or wherever the itch happens to be. The first is induced hypersensitivity. When fleas or head lice are mentioned, the brain increases the sensitivity of its monitoring of the skin and any slight sensation – feelings your skin is detecting all the time – are over-amplified and interpreted as an itch. (As I type this I am genuinely fighting the urge to scratch itches. You may well be too.)

It has also been demonstrated in trials that, like yawning, itching can be visually transmitted. If you see someone else scratching – or even a video of someone else scratching – it pushes the brain into overdrive and the tiniest changes in your skin, which are happening all the time due to air movement, body movements and so on – are amplified, prompting the urge to scratch. The same thing happens with monkeys – it's thought it might be part of the grooming urge to remove parasites among groups of primates.

The second possibility arises when you get an itch when you are trying to keep still. What seems to be happening here is a result of the way your brain manages consciousness. The majority of the things your brain does in interaction with your body are unconscious. Your conscious mind only deals with a small percentage of its role. Generally speaking, if you are actively thinking about something, you won't notice a minor itch. But as soon as you try to be still and receptive, the conscious mind is relatively unoccupied and the itch comes to the fore.

Lottery

It's traditional for scientists and mathematicians to point out that entering a lottery is a waste of time – but they are using the wrong branch of science.

From the mathematician's viewpoint you have to be an idiot to enter a lottery like the UK's National Lottery Lotto draw. Why? Because your chance of winning the jackpot is tiny. To be precise, it's 13,983,815 to one. By comparison, the chance of being struck by lightning in the UK is about 1,000,000 to one. (This isn't quite true. That's just the population of the UK divided by the number of people hit by lightning. But people who are regularly in places where they are likely to get hit by lightning are much more likely to be hit, the rest of us much less so. Still, the two probabilities are comparable in size.)

However, we should be employing psychology rather than statistics. Entering the lottery is not just about probability – it is an entertainment, not a very risky investment. You might as well say that you pay money for nothing when you go to a theme park or the opera, because you are no different after you go than you were before. With the lottery, the stake is one that most of us can afford to lose, and we play as much for the thrill of anticipation as for the highly unlikely outcome of winning a jackpot.

In the UK there is an alternative to the lottery that is arguably better – premium bonds. Here your 'tickets' are entered in a monthly draw that can win a million pounds plus lots of smaller prizes, but the two big advantages are that once you have a ticket it is entered every month, and you can cash in that ticket at any time for the amount you paid for it. All you lose is the interest you would have got had it been invested in a bank. Seen as a pure investment, premium bonds provide a relatively poor rate of interest (1.3 per cent per annum at the time of writing – but still far better than the lottery, which is the equivalent of a sizeable negative rate). However, including the excitement factor, premium bonds are not a bad way to tuck money away, especially when interest rates are low.

Queuing

Being in a queue is hardly fun, but we do it all the time, so it's certainly quite interesting to give some thought to the science of the queue and how we can do it better.

We spend a surprising amount of time in queues, and it's something most of us would rather not do – so why not improve them?

Sometimes, those involved in making us queue do try to make the experience better. For instance, the better theme parks, which often subject us to some of the longest queues we regularly face on foot, try to put entertainment in the queue – but most organisations that put us in a queue leave us to our own devices. It's well worth ensuring you are equipped to keep yourself busy. A smartphone or a tablet can transform the queuing experience, though you may get some dirty looks from other members of the queue, who often think that this is antisocial.

The biggest contribution that businesses have made to the queuing experience is the multi-queue. At one time every business with multiple desks/gates/checkouts had individual queues for each server. But the science of queuing was hugely advanced by being able to make computer models of queues (this was something I did while working at British Airways), and it became clear that far fewer people would have a long wait in a queue if you had a single queue feeding all the servers.

To begin with, customers resisted this, feeling that the multi-queue would make them wait longer because there were more people in front of them, but now they are universally accepted. There's an interesting social division between the London area and the rest of the UK on this. In London, unless a multi-queue is forced on people, they will go with their gut rather than their head, and charge straight for a specific server. But in much of the rest of the country, multi-queues form automatically – in fast food outlets and even in front of a bank of cash machines, individuals will wait a few paces back from the servers and allocate themselves to the next available place.

Supermarkets generally resist the multi-queue, resulting in that terrible decision of which till to go to, where you immediately find that your queue is going more slowly than anyone else's. This probably isn't true, but your perception will make it seem slower. When those simulations are run, there is no overall benefit from switching queues, unless you switch to a newly opened queue. This provides a useful tactic in choosing which till to queue at in a supermarket.

Don't go for either end, because these tend to be particularly popular and because they have only one neighbour. Instead, go for a till that has closed tills either side of it – that way, if either opens you can nip in ahead of the rest.

In principle, supermarkets could adopt multi-queue systems, but it would require them to leave a large space between the tills and the shelves to accommodate a long queue of trolleys. (A few smaller supermarkets do this, but it is uncommon.) However, the advent of the self-service till has brought them in. Automated tills tend to divide opinion. Some are convinced (in part by TV jokes) that the tills are slow and cumbersome. But most of the time they are faster than a traditional till, and they usually have a multi-queue. Forget the fear of 'unexpected item in the bagging area' – they really do speed up your shop.

One particularly interesting (and frustrating) queue is the kind of traffic jam you find on motorways, usually when you are in a hurry. Again, these queues have been widely studied with simulations and have some strange properties. It often seems that when you come to the end of the queue there was nothing to cause it. This is because once the queue has been established it will carry on long after the original problem has been removed. The 'hold up' point travels backwards up the motorway like a wave.

Another feature of motorway queues is the way that whenever you decide to change lane because yours is going slower, the lane you move into slows down. This strange behaviour happens because a queue on a road is not like any traditional queue – you can enter it at any point. If a lane seems to be moving more quickly, then people will start moving into it all along its length. Suddenly there are far more people in that lane, and so it slows down.

You will also find that the fastest moving lane is often the one that is blocked. This is because this is the only lane that everyone has to get out of at some point. So traffic is constantly moving from that lane to the others, slowing down adjacent lanes and speeding up the blocked one.

Recorded music

**If you are inclined to pay large sums of money
for the latest high definition sound system,
make sure you know what you are buying.**

Music enthusiasts are prepared to pay a lot of money to get the best sound reproduction, and music equipment makers are happy to take their money. But the shoppers don't always get a lot for their money.

A while ago there was an attempt to supplant CDs with Super Audio CD (SACD), which had more information stored in the digital format so was thought to be 'better'. Unfortunately, the CD already spans further than the range the human ear can cope with, and in blind trials experts were not able to tell the difference because the extra sound range was totally wasted.

There is more concern about MP3s and other digital music files as these compress the sound, losing fine detail in exchange for taking up less storage space. The earliest examples were quite poor, but for most ears, current levels of compression are as much as can be distinguished, and those attempting to sell very expensive high rate audio players are again looking to make money from those who are prepared to pay without really worrying about whether there is any advantage. (The same goes for companies selling audio leads at 50 times the price of a standard one. There is only so much advantage you can get in a wire.)

So, the next time you are tempted to pay hundreds of pounds for the latest trendy headphones, or to get a next generation music player, make sure you know just what benefit you will get from paying for the ultimate audio experience.

Toilet roll perforations

Surprisingly, that curse of the toilet roll – perforations getting out of sync – is not what it seems.

We've all been there. Every time you attempt to tear a section off a toilet roll, it doesn't work properly because the perforations aren't lined up in the different layers of a multi-ply toilet tissue, and you wonder why they perforate each layer separately rather than doing them all together, which would prevent this happening.

Actually they do perforate them all together – the problem to consider is how the paper gets to its current state. There's a bit of a parallel in the way that the sound can get out of sync when you are watching some kinds of video, particularly streaming sources. It's not that it was recorded out of sync, but the playback mechanism has had a glitch. So, it is usually fixed by stopping the playback and starting it again. It's the same with the toilet roll.

Usually what has happened when the perforations get out of line is that someone has unwrapped an extra rotation of the outer layer. As there aren't an exact number of sheets in the circumference of the roll, this puts the layers out of sync. All you need to do to restore harmony is to unwrap or re-wrap a rotation of the outer layer until it all lines up once more.

UFOs

UFOs (or unidentified flying objects) are a subject that causes constant fascination, but what is the scientific evidence, and has there been a government cover-up?

There is no doubt that UFOs exist. But this isn't as impressive as it sounds, since all a UFO has to be is something that flies that we haven't identified. That doesn't make it a flying saucer or an alien spaceship. And as yet, despite all the hype, there is no good scientific evidence of the existence of alien spaceships – which is surprising given the number of reported sightings.

For instance, where are all the good flying saucer pictures? You might think that there are lots, but most flying saucer pictures are really, really bad or look fake (or both). I know a little about such fakes as I went through a phase of making them in my teens. Broadly, there were two kinds before digital photo editing. A relatively detailed model, suspended against the sky with fishing line, which typically had to be a little out of focus to cover up that this is what it was, or a hubcap or metal plate, thrown high in the sky, frisbee style.

The trouble with this second approach is that the thrown object usually travelled at an angle making its flight look unrealistic. And fascinatingly, several of the 'UFOs' shown in books have exactly the same problem – flying at a weird angle, just like my hubcap. It is also amazing that many books include a famous picture of UFOs over the night-time Capitol building in Washington. It does look impressive: a formation of flying lights hovering over the dome of this imposing structure. That's what you see in the book. But if you take a look at the uncropped version of the photo, there are a series of street lights on the steps in front of the building, in *exact* mirror formation to the 'UFOs'. The UFOs are just the camera producing a reflection of the bright lights on the dark sky.

The reason UFO photos were so universally awful used to be explained as being because the majority of the people who claimed to have seen UFOs didn't have cameras with them. So it was just that one in 1,000 times someone did have a camera that we got the shots, and there were sufficiently few that most were rubbish. Only it's not like that anymore. These days most of us carry both a camera and a video camera on our phones. So why haven't we seen a sudden burst of vast quantities of good photographic/video evidence of UFOs and little grey men?

Sadly, the answer seems straightforward. Because they were never there in the first place. And that makes a lot of sense from a scientific viewpoint – because the distances in space are so great. The nearest star apart from the Sun is around four light years away, and the nearest so far found that might have a habitable planet is around twenty light years away, with most much further. (A light year is the distance light travels in one year, quite a distance when you consider light moves at 300,000 kilometres a second.)

The fastest a human being has ever gone was on Apollo 10, where they travelled at 39,896 kilometres per hour with respect to the Earth – around 0.000037 times the speed of light. At this speed, it would take around 5,400 years to reach that nearest likely habitable planet. Of course, alien technology could be faster than ours – but there are major physical problems involved in getting anywhere near quick enough to cross interstellar space. Throw in the sheer size of the universe – our galaxy alone contains around 300 billion stars – and the chances of aliens turning up at our planet, particularly in the numbers reported, are pretty small.

By far the most likely explanation is our imaginations.

Weather forecasts

**There is nothing that can make or break a
fun day like the weather. So why, with all
the money we throw at meteorologists,
are weather forecasts so bad?**

Weather forecasts have improved hugely over the last twenty years
– but they do sometimes get it wrong, and we need to face up to
the fact that this is inevitable. This is because weather is the kind of
system described by mathematicians as 'chaotic'. This doesn't mean
that it is random, but rather that it is so complex, with so many
factors influencing the outcome, that tiny differences in how things
start out can make huge differences a little way down the line. As
the measurements meteorologists make can never be perfectly exact
(between whether the temperature is 20.001 degrees or 20.002, for
instance), there will always be room for error.

The reason that forecasts have got so much better since the forecasters missed the great storm that devastated the UK in 1987, is
that they no longer try to produce a single definitive forecast. With
the computer power at their disposal they can make many different
forecasts with subtly different variants on the starting state of the
weather. That way they can then give much better information like
the chance of rainfall (e.g. a 60 per cent chance of rain), by averaging
across the predictions of the different models.

So, looking forward a day or two, weather forecasts are now
pretty accurate. Why then is it that we still get the feeling that they
are often wrong? The media are largely to blame. Time and again the
more irresponsible newspapers will splash headlines along the lines
of 'Expect the hottest summer in 50 years' or 'Torrential downpours
expected' or 'Brace yourself for a long, snowy winter'. And mostly
these predictions are useless.

The trouble is that because of the chaotic nature of the weather, any attempt to forecast more than five days ahead is fairly doubtful, and more than ten days is futile. Studies have shown that you have a better chance of predicting the weather a month ahead by simply describing what it is typically like at that time of year in a specific location than by using weather forecasting models. So, whenever you hear predictions in April about what the summer is going to be like, don't blame the forecasters. They know they can't say anything sensible. Blame the media (and the rest of us) for demanding an answer, however wrong it might be.

Wine

Wines vary hugely in price – but is that price any indication of whether a wine is better or not? Making the right choice could save you hundreds of pounds.

There are several entries earlier in the book on the effects of alcohol and wine on health. Here we are considering something quite different – whether the pleasure we get from wine is related to what we pay for it. It's possible to pay anything from around £1.50 (in France) to £1,000+ for a bottle of wine. A standard bottle contains 750ml of the stuff, so the question is, does the extra money bring you a proportionate amount of extra enjoyment, or is it all hype?

There can be no doubt that there are rough wines – the kind that give the impression they would make a good paint stripper. But by the time we reach a basic standard – perhaps at around the £3.50 a bottle level (at the time of writing) – wine tends to be perfectly palatable, and after that we are looking for subtleties.

Some facts are pretty well known when it comes to making the choice. Given the option of a decent everyday supermarket red and a top-of-the range, Premier Cru claret, most people prefer the supermarket red in a blind tasting. Experts won't, because they are looking for something very specific, including multiple layers of taste that the vast majority of ordinary wine buyers will never notice. So, unless you are that wine expert, lashing out on a bottle that may well cost 100 times as much as a basic one is absolutely silly.

Even wine enthusiasts can be fooled. In *Think Like a Freak*, Steven Levitt and Stephen Dubner recount an experiment where they performed a blind tasting at a Harvard society with an impressive cellar and where many of those present considered themselves wine connoisseurs. They weren't professionals, but they were people who 'knew their wine'. This wasn't a big enough experiment to be definitive, but it was indicative that given three expensive wines and one cheap one, all the wines were rated very similarly. In fact, the biggest distinction was considered to be between two decanters that contained the same wine.

In reality, we are hugely swayed by labels and presentation. If you really enjoy wine, but don't want to pay over the odds, it is worth organising yourself a little blind tasting. Get a range of wines from a cheap offer to maybe twice what you normally pay. Label glasses on the bottom and pour out the wines, then put the bottles away. Get someone else to shuffle the glasses so you genuinely don't know which is which. Then rank the wines in order of taste. If the only ones you like turn out to be the expensive ones – fine. But the chances are you will also like some cheap ones and you can identify what works for you. Doing the test might set you back a bit, but if you would otherwise spend the rest of your life buying a wine that's more expensive, it will save you huge amounts.

One last pointer: always be aware how much you are influenced by branding, whether at the level of a particular label or a region. I always celebrate with sparkling Saumur rather than champagne.

It is produced by the same method, is also available in Brut if you like your champagne nice and dry, and to me tastes nicer than most champagnes – yet costs around one third of the price. A double bonus.

LINKS:
- **Alcohol** – page 5
- Red wine and health – see **Red wine** – page 87

Yet more

Remember there's even more Science for Life available online, keeping up with the latest science news.

Visit **www.scienceforlife.info**